Applied Petroleum Geochemistry

Applied Petroleum Geochemistry

Editor

Gabriel Brown

AURIS Reference

AURIS REFERENCE LTD.

London, UK

Applied Petroleum Geochemistry

Published by
Auris Reference Ltd., UK
www.aurisreference.com

ISBN: 978-1-78154-336-8

Editor: Gabriel Brown

Printed in UK

10 9 8 7 6 5 4 3 2 1

British Library Cataloguing in Publication Data
A CIP record for this book is available from the British Library

Reasonable efforts have been made to publish reliable data and information, but the authors, editors, and the publisher cannot assume responsibility for the legality of all materials or the consequences of their use. The authors, editors, and the publisher have attempted to trace the copyright holders of all materials in this publication and express regret to copyright holders if permission to publish has not been obtained. If any copyright material has not been acknowledged, let us know so we may rectify in any future reprint.

For information about Auris Reference Ltd and its publications, visit our website at www.aurisreference.com

Contents

Preface

Petroleum is a naturally occurring, smelly, yellow-to-black liquid consisting of a complex mixture of hydrocarbons of various molecular weights and other liquid organic compounds, that are found in geologic formations beneath the Earth's surface. The name Petroleum covers both naturally occurring unprocessed crude oils and petroleum products that are made up of refined crude oil. A fossil fuel, petroleum is formed when large quantities of dead organisms, usually zooplankton and algae, are buried underneath sedimentary rock and undergo intense heat and pressure.

Petroleum geochemistry is the application of chemical principles to the study of the origin, migration, accumulation, and alteration of petroleum. Petroleum is generally considered oil and natural gases having various compounds composed of primarily hydrogen and carbon. They are usually generated from the decomposition and/or thermal maturation of organic matter. The organic matter originated from plants and algae. The organic matter is deposited after the death of the plant in sediments, where after considerable time, heat, and pressure the compounds in the plants and algae are altered to oil, gas, and kerogen. Kerogen can be thought of as the remaining solid material of the plant. The sediment - usually clay and/or calcareous (lime) ooze, hardens during this alteration process into rock i.e. shale and/or limestone. The shale or limestone rock containing the organic matter is called the source rock because it is the source, having generated the petroleum. Petroleum is not as easy to find as it used to be. In order to locate and develop reserves efficiently, it's vital that geologists and geophysicists understand the geological processes that affect a reservoir rock and the oil that is trapped within it.

This book is written for: petroleum geologists and geophysicists; managers who should integrate the impact of geochemistry in exploration decision-making; specialized geochemists who need an accurate panorama of other aspects of geochemistry; university professors and students in petroleum geology.

—*Editor*

Chapter 1

Principles of Geochemistry

The field of geochemistry involves study of the chemical composition of the Earth and other planets, chemical processes and reactions that govern the composition of rocks, water, and soils, and the cycles of matter and energy that transport the Earth's chemical components in time and space, and their interaction with the hydrosphere and the atmosphere.

Some subsets of geochemistry are:

Isotope Geochemistry: Determination of the relative and absolute concentrations of the elements and their isotopes in the earth and on earth's surface.

Figure: *Grand Canyon*

Examination of the distribution and movements of elements in different parts of the earth (crust, mantle, hydrosphere etc.) and in minerals with the goal to determine the underlying system of distribution and movement.

Cosmochemistry: Analysis of the distribution of elements and their isotopes in the cosmos.

Biogeochemistry: Field of study focusing on the effect of life on the chemistry of the earth.

Organic Geochemistry: A study of the role of processes and compounds that are derived from living or once-living organisms.

Aqueous Geochemistry: Understanding the role of various elements in watersheds, including copper, sulphur, mercury, and how elemental fluxes are exchanged through atmospheric-terrestrial-aquatic interactions. Regional, environmental and exploration geochemistry: Applications to environmental, hydrological and mineral exploration studies.

Victor Goldschmidt is considered by most to be the father of modern geochemistry and the ideas of the subject were formed by him in a series of publications from 1922 under the title 'Geochemische Verteilungsgesetze der Elemente' (geochemical laws of distribution of the elements).

Chemical Characteristics

The more common rock constituents are nearly all oxides; chlorides, sulphides and fluorides are the only important exceptions to this and their total amount in any rock is usually much less than 1%. F. W. Clarke has calculated that a little more than 47% of the Earth's crust consists of oxygen. It occurs principally in combination as oxides, of which the chief are silica, alumina, iron oxides, and various carbonates (calcium carbonate, magnesium carbonate, sodium carbonate, and potassium carbonate).

The silica functions principally as an acid, forming silicates, and all the commonest minerals of igneous rocks are of this nature. From a computation based on 1672 analyses of numerous kinds of rocks Clarke arrived at the following as the average percentage composition: SiO_2=59.71, Al_2O_3=15.41, Fe_2O_3=2.63, FeO=3.52, MgO=4.36, CaO=4.90, Na_2O=3.55, K_2O=2.80, H_2O=1.52, TiO_2=0.60, P_2O_5=0.22, total 99.22%). All the other constituents occur only in very small quantities, usually much less than 1%.

These oxides combine in a haphazard way. For example, potash (potassium carbonate) and soda (sodium carbonate) combine to produce feldspars. In some cases they may take other forms, such as nepheline, leucite, and muscovite, but in the great majority of instances they are found as feldspar. Phosphoric acid with lime (calcium carbonate) forms apatite. Titanium dioxide with ferrous oxide gives rise to ilmenite. Part of the lime forms lime feldspar. Magnesium carbonate and iron oxides with silica crystallize as olivine or enstatite, or with alumina and lime form the complex ferro-magnesian silicates of which the pyroxenes, amphiboles, and biotites are the chief. Any excess of silica above what is required to neutralize the bases will separate out as quartz; excess of alumina crystallizes as corundum. These must be regarded only as general tendencies. It is possible, by rock analysis, to say approximately what minerals the rock contains, but there are numerous exceptions to any rule.

Mineral Constitution

Hence we may say that except in acid or siliceous rocks containing 66% of silica and over, quartz will not be abundant. In basic rocks (containing 20% of silica or less) it is rare and accidental. If magnesia and iron be above the average while silica is low, olivine may be expected; where silica is present in greater quantity over ferro-magnesian minerals, such as augite, hornblende, enstatite or biotite, occur rather than olivine.

Unless potash is high and silica relatively low, leucite will not be present, for leucite does not occur with free quartz. Nepheline, likewise, is usually found in rocks with much soda and comparatively little silica. With high alkalis, soda-bearing pyroxenes and amphiboles may be present. The lower the percentage of silica and the alkalis, the greater is the prevalence of calcium feldspar as contracted with soda or potash feldspar. Clarke has calculated the relative abundance of the principal rock-forming minerals with the following results: apatite=0.6, titanium minerals=1.5, quartz=12.0, feldspars=59.5, biotite=3.8, hornblende and pyroxene=16.8, total=94.2%. This, however, can only be a rough approximation.

The other determining factor, namely the physical conditions attending consolidation, plays on the whole a smaller part, yet is by no means negligible, as a few instances will prove. Certain minerals are practically confined to deep-seated intrusive rocks, e.g., microcline, muscovite, diallage. Leucite is very rare in plutonic masses; many minerals have special peculiarities in microscopic character according to whether they crystallized in depth or near the surface, e.g., hypersthene, orthoclase, quartz.

There are some curious instances of rocks having the same chemical composition, but consisting of entirely different minerals, e.g., the hornblendite of Gran, in Norway, which contains only hornblende, has the same composition as some of the camptonites of the same locality that contain feldspar and hornblende of a different variety. In this connection we may repeat what has been said above about the corrosion of porphyritic minerals in igneous rocks.

In rhyolites and trachytes, early crystals of hornblende and biotite may be found in great numbers partially converted into augite and magnetite. Hornblende and biotite were stable under the pressures and other conditions below the surface, but unstable at higher levels. In the ground-mass of these rocks, augite is almost universally present. But the plutonic representatives of the same magma, granite and syenite contain biotite and hornblende far more commonly than augite.

Acid, Intermediate and Basic Igneous Rocks

Those rocks that contain the most silica, and on crystallizing yield free quartz, form a group generally designated the "acid" rocks. Those again that contain least silica and most magnesia and iron, so that quartz is absent while olivine is usually abundant, form the "basic" group. The "intermediate" rocks include those characterized by the general absence of both quartz and olivine.

An important subdivision of these contains a very high percentage of alkalis, especially soda, and consequently has minerals such as nepheline and leucite not common in other rocks. It is often separated from the others as the "alkali" or "soda" rocks, and there is a corresponding series of basic rocks. Lastly a small sub-group rich in olivine and without feldspar has been called the "ultrabasic" rocks. They have very low percentages of silica but much iron and magnesia.

Except these last, practically all rocks contain felspars or feldspathoid minerals. In the acid rocks the common feldspars are orthoclase, perthite, microcline, and oligoclase—all having much silica and alkalis. In the basic rocks labradorite, anorthite and bytownite prevail, being rich in lime and poor in silica, potash and soda. Augite is the commonest ferro-magnesian of the basic rocks, but biotite and hornblende are on the whole more frequent in the acid.

Figure: *Basic Igneous Rocks*

Rocks that contain leucite or nepheline, either partly or wholly replacing felspar, are not included in this table. They are essentially of intermediate or of basic character. We might in consequence regard them as varieties of syenite, diorite, gabbro, etc., in which feldspathoid minerals occur, and indeed there are many transitions between syenites of ordinary type and nepheline — or leucite — syenite, and between gabbro or dolerite and theralite or essexite. But, as many minerals develop in these "alkali" rocks that are uncommon elsewhere, it is convenient in a purely formal classification like that outlined here to treat the whole assemblage as a distinct series.

This classification is based essentially on the mineralogical constitution of the igneous rocks. Any chemical distinctions between the different groups, though implied, are relegated to a subordinate position. It is admittedly artificial by it has grown up with the grown of the science and is still adopted as the basis on which more minute subdivisions are erected.

The subdivisions are by no means of equal value. The syenites, for example, and the peridotites, are far less important than the granites, diorites and gabbros. Moreover, the effusive andesites do not always correspond to the plutonic diorites but partly also to the gabbros. As the different kinds of rock, regarded as aggregates of minerals, pass gradually into one another, transitional types are very common and are often so important as to receive special names. The quartz-syenites and nordmarkites may be interposed between granite and syenite, the tonalites and adamellites between granite and diorite, the monzoaites between syenite and diorite, norites and hyperites between diorite and gabbro, and so on.

An atom is composed of a nucleus of protons and neutrons surrounded by an electron cloud. Theoretically, electrons may be found at any distance from the nucleus, although they preferentially rotate around low-energy orbits or *levels*.

Within a single level, various *sublevels* can be distinguished. [The term "level" corresponds to "electron shell" in the Bohr model. The terminological analogy is: shell K = level I (n = 1); shell L = level II (n = 2); shell M = level III (n = 3); shell N = level IV (n = 4); and so on.] Electron levels are established according to four quantum numbers:

1. The principal quantum number (n) is related to the energy of the electron levels. An increase in quantum number n corresponds to an increase in the distance of the electron from the nucleus.
2. The orbital or "azimuthal" quantum number (l) defines "form" and indicates which sublevel is occupied by the electron. It assumes integer values between 0 and n–1.
3. The magnetic quantum number (m_l) indicates which orbital of a given sublevel is occupied by the electron. For a given orbital, quantum number l assumes integer values between "1 and +1. An electron with a magnetic quantum number m_l subjected to a magnetic field B increases its energy by a term $E = \mu_b m_l$ B, where μ_b is Bohr's magneton (Zeeman effect).
4. The spin quantum number (m_s) gives the projection of the intrinsic angular momentum of the electron, with values +½ and –½. Two oppositely oriented spins on the same orbital are graphically represented by arrows in opposite directions (↓↑).

The development of quantum mechanics has led to the formulation of various principles governing the structure of atoms:

1. The Aufbau or "build-up" principle states that electrons occupy orbitals of progressively increasing energies.
2. The Pauli Exclusion Principle states that no more than two electrons may occupy the same orbital, and they must have opposite spins. Based on this principle, 2n is the maximum number of electrons compatible with a given level.
3. Hund's rule states that, before two opposite-spin electrons can occupy the same orbital of a given sublevel, all the other orbitals of the same sublevel must have been occupied by single electrons of equal spin.

Wavelike Behaviour of Electrons

Like any particle, an electron has a definite mass, corresponding to

$$\frac{1}{1850}$$

of the mass of the hydrogen atom. Nevertheless, experiments with electron beams accelerated by known voltage differences show that a wavelength λ can be associated with the mass of the electron, so that

$$\lambda = \frac{h}{mv}, \tag{1}$$

where *h* is Planck's constant and v is velocity. Equation 1, which is one of the fundamental results of quantum mechanics, establishes the dual nature of electrons as particles and wavelike elements.

The propagation of a wave in one dimension is obtained by the partial derivative of the wave function

$$\psi = \psi_o \sin 2\pi\left(\frac{x}{\lambda} - vt\right). \tag{2}$$

where v is frequency (v = v/λ) and *t* is time. Function 2 is periodic in *t* when *x* is constant (each point vibrates) and periodic in *x* when *t* is constant (instantaneous picture of a wave). Partial differentiation of function 2 on *x* results in

$$\frac{\partial^2 \psi}{\partial x^2} = -\frac{4\pi}{\lambda^2}\psi. \tag{3}$$

Principle Quantum Number, *n*	**AngularMomentum uantum Number, l**	**Magnetic Quantum Number, ml**	**Electron Summary**
1	0 (s)	0	1 1s orbital, 2 electrons
2	0 (s)	0	1 2s orbital, 2 electrons
	1 (p)	1, 0, -1	3 2p orbitals, 6 electrons
3	0 (s)	0	1 3s orbitals, 2 electrons
	1 (p)	1, 0, -1	3 3p orbitals, 6 electrons
	2 (d)	2, 1, 0, -1, -2	5 3d orbitals, 10 electrons
4	0 (s)	0	1 4s orbitals, 2 electrons
	1 (p)	1, 0, -1	3 4p orbitals, 6 electrons
	2 (d)	2, 1, 0, -1, -2	5 4d orbitals, 10 electrons
	3 (f)	-3, -2, -1, 0, 1, 2, 3	7 4f orbitals, 14 electrons

The three-dimensional analogue of function 2 is

$$\psi = \psi_0 \sin 2\pi\left[\frac{(x^2 + y^2 + z^2)^{1/2}}{\lambda} - vt\right]. \tag{4}$$

and it can be shown that partial derivation of equation 4 over the three dimensions results in

$$\frac{\partial^2\psi}{\partial x^2}+\frac{\partial^2\psi}{\partial y^2}+\frac{\partial^2\psi}{\partial z^2}+-\frac{4\pi}{\lambda^2}\psi. \qquad (5)$$

Let us now consider a particle with mass m, total energy E, and potential energy Φ. The kinetic energy of particle E_k will be

No.	***Element***	***1***	***2***	***3***	***4***	***5***	***6***	***7***
1	Hydrogen	1						
2	Helium	2						
3	Lithium	2	1					
4	Beryllium	2	2					
5	Boron	2	3					
6	Carbon	2	4					
7	Nitrogen	2	5					
8	Oxygen	2	6					
9	Fluorine	2	7					
10	Neon	2	8					
11	Sodium	2	8	1				
12	Magnesium	2	8	2				
13	Aluminium	2	8	3				
14	Silicon	2	8	4				
15	Phosphorus	2	8	5				
16	Sulfur	2	8	6				
17	Chlorine	2	8	7				
18	Argon	2	8	8				
19	Potassium	2	8	8	1			
20	Calcium	2	8	8	2			
21	Scandium	2	8	9	2			
22	Titanium	2	8	10	2			
23	Vanadium	2	8	11	2			
24	Chromium	2	8	13	1			
25	Manganese	2	8	13	2			
26	Iron	2	8	14	2			
27	Cobalt	2	9	15	1			
28	Nickel	2	8	17	1			

Contd...

No.	Element	1	2	3	4	5	6	7
29	Copper	2	8	18	1			
30	Zinc	2	8	18	2			
31	Gallium	2	8	18	3			
32	Germanium	2	8	18	4			
33	Arsenic	2	8	18	5			
34	Selenium	2	8	18	6			
35	Bromine	2	8	18	7			
36	Krypton	2	8	18	8			
37	Rubidium	2	8	18	8	1		
38	Strontium	2	8	18	8	2		
39	Yttrium	2	8	18	9	2		
40	Zirconium	2	8	18	10	2		
41	Niobium	2	8	18	12	1		
42	Molybdenum	2	8	18	13	1		
43	Technetium	2	8	18	13	2		
44	Ruthenium	2	8	18	15	1		
45	Rhodium	2	8	18	16	1		
46	Palladium	2	8	18	18	0		
47	Silver	2	8	18	18	1		
48	Cadmium	2	8	18	18	2		
49	Indium	2	8	18	18	3		
50	Tin	2	8	18	18	4		
51	Antimony	2	8	18	18	5		
52	Tellurium	2	8	18	18	6		
53	Iodine	2	8	18	18	7		
54	Xenon	2	8	18	18	8		
55	Caesium	2	8	18	18	8	1	
56	Barium	2	8	18	18	8	2	
57	Lanthanum	2	8	18	18	9	2	
58	Cerium	2	8	18	19	9	2	
59	Praseodymium	2	8	18	21	8	2	
60	Neodymium	2	8	18	22	8	2	
61	Promethium	2	8	18	23	8	2	
62	Samarium	2	8	18	24	8	2	

Contd...

No.	***Element***	***1***	***2***	***3***	***4***	***5***	***6***	***7***
63	Europium	2	8	18	25	8	2	
64	Gadolinium	2	8	18	25	9	2	
65	Terbium	2	8	18	27	8	2	
66	Dysprosium	2	8	18	28	8	2	
67	Holmium	2	8	18	29	8	2	
68	Erbium	2	8	18	30	8	2	
69	Thulium	2	8	18	31	8	2	
70	Ytterbium	2	8	18	32	8	2	
71	Lutetium	2	8	18	32	9	2	
72	Hafnium	2	8	18	32	10	2	
73	Tantalum	2	8	18	32	11	2	
74	Tungsten	2	8	18	32	12	2	
75	Rhenium	2	8	18	32	13	2	
76	Osmium	2	8	18	32	14	2	
77	Iridium	2	8	18	32	15	2	
78	Platinum	2	8	18	32	17	1	
79	Gold	2	8	18	32	18	1	
80	Mercury	2	8	18	32	18	2	
81	Thallium	2	8	18	32	18	3	
82	Lead	2	8	18	32	18	4	
83	Bismuth	2	8	18	32	18	5	
84	Polonium	2	8	18	32	18	6	
85	Astatine	2	8	18	32	18	7	
86	Radon	2	8	18	32	18	8	
87	Francium	2	8	18	32	18	8	1
88	Radium	2	8	18	32	18	8	2
89	Actinium	2	8	18	32	18	9	2
90	Thorium	2	8	18	32	18	10	2
91	Protactinium	2	8	18	32	20	9	2
92	Uranium	2	8	18	32	21	9	2
93	Neptunium	2	8	18	32	22	9	2
94	Plutonium	2	8	18	32	24	8	2
95	Americium	2	8	18	32	25	8	2

Contd...

No.	*Element*	*1*	*2*	*3*	*4*	*5*	*6*	*7*
96	Curium	2	8	18	32	25	9	2
97	Berkelium	2	8	18	32	27	8	2
98	Californium	2	8	18	32	28	8	2
99	Einsteinium	2	8	18	32	29	8	2
100	Fermium	2	8	18	32	30	8	2
101	Mendelevium	2	8	18	32	31	8	2
102	Nobelium	2	8	18	32	32	8	2
103	Lawrencium	2	8	18	32	32	8	3
104	Rutherfordium	2	8	18	32	32	10	2
105	Dubnium	2	8	18	32	32	11	2
106	Seaborgium	2	8	18	32	32	12	2
107	Bohrium	2	8	18	32	32	13	2
108	Hassium	2	8	18	32	32	14	2
109	Meitnerium	2	8	18	32	32	15	2
110	Darmstadtium	2	8	18	32	32	16	2
111	Roentgenium	2	8	18	32	32	17	2
112	Copernicium	2	8	18	32	32	18	2
113	Ununtrium	2	8	18	32	32	18	3
114	Ununquadium	2	8	18	32	32	18	4
115	Ununpentium	2	8	18	32	32	18	5
116	Ununhexium	2	8	18	32	32	18	6
117	Ununseptium	2	8	18	32	32	18	7
118	Ununoctium	2	8	18	32	32	18	8

$$E_K = \frac{1}{2}mv^2 = E - \Phi. \tag{6}$$

From equation 6 we obtain

$$m^2v^2 = 2m(E - \Phi). \tag{7}$$

which, when combined with equation 1, gives

$$\frac{1}{\lambda^2} = \frac{2\pi(E - \Phi)}{h^2} \tag{8}$$

Substituting the right side of equation 8 into wave equations 3 and 5 leads, respectively, to

$$\frac{d^2\psi}{dx^2}+\frac{8\pi^2 m}{h^2}(E-\Phi)\psi=0 \tag{9}$$

and

$$\nabla^2\psi+\frac{8\pi^2 m}{h^2}(E-\Phi)\psi=0 \tag{10}$$

where Δ^2 is the Laplacian operator

$$\nabla^2=\frac{\partial^2}{\partial x^2}+\frac{\partial^2}{\partial y^2}+\frac{\partial^2}{\partial z^2} \tag{11}$$

Equation 9 is the *Schrödinger equation for one-dimensional stationary states,* and equation 10 is the tridimensional form. Differential equations 9 and 10 possess finite single-valued solutions, within the domain of squared integrable functions, only for definite values of energy (*eigenvalues*). For vibrational systems, such values are given by

$$E_{(n)}=\left(n+\frac{1}{2}\right)hv, \tag{12}$$

where *n* is the *vibrational quantum number.*

Applying to equation 9 the potential of the *monodimensional harmonic oscillator* (i.e., a particle constrained to a certain equilibrium position by a force that is proportional to its displacement from the origin), we obtain

$$\frac{d^2\psi_{(x)}}{dx^2}+\frac{8\pi^2}{h^2}\left(E-\frac{K_F x^2}{2}\right)\psi_{(x)}=0. \tag{13}$$

If position *x* of the particle corresponds to its displacement from the origin, the restoring force acting on the particle by virtue of the harmonic potential is given by *Hooke's law:*

$$F=-K_F x, \tag{14}$$

where K_F is the force constant.

Because, in a conservative field, the force is linked to the gradient of the potential by

$$F=-\frac{d\Phi_{(x)}}{dx}, \tag{15}$$

by integration of equation 15 it follows that

$$\Phi_{(x)}=\frac{K_F x^2}{2}, \tag{16}$$

which, substituted into equation 9, leads to equation 13.

The eigenvalues of the energy that satisfy equation 1.13 are given by

$$E_{(n)} = \frac{K_F h^2}{4\pi^2 m}\left(n + \frac{1}{2}\right). \tag{17}$$

We will later adopt function 17 when treating the vibrational energy of crystalline substances and its relationship to entropy and heat capacity.

A more detailed account of the Schrödinger equation can be found in physical chemistry textbooks such as those of Hinshelwood (1951) and Atkins (1978), or in more specialised texts such as that of Hirschfelder et al. (1954). An excellent review of the applications of quantum mechanics to geochemistry has recently been proposed by Tossell and Vaughan (1992).

Periodic Properties of the Elements

The periodic law, discovered by Dmitri Mendeleev, establishes "an harmonic periodicity of properties of chemical individuals, dependent on their masses". To highlight the existence of periodic properties, the various elements are arranged, according to their increasing masses, in groups (8), subgroups (8), and periods (7). A given group comprises elements with similar properties. The periodicity of elemental properties is easily understood if we recall the laws governing the occupancy of atomic orbitals. The periodic table shows the electron structures of the various elements. For the sake of brevity, the structure of the inner shells is indicated by referring to that of the preceding inert gas. An inert gas displays no chemical reactivity (i.e., no tendency to enter into other states of combination). This "stable" configuration corresponds to completion of all sublevels of the principal quantum number *n*. As we can see, a period begins with the occupancy of a new level; elements in this position are ready to lose one electron, with the formation of a univalent positive ion. At the end of the period, immediately preceding the inert gas, the atoms lack a single electron to complete the stable configuration. Such atoms tend to be transformed into univalent negative ions by electron capture. The number of electrons that complete the period is readily calculated from the following summation:

$$\sum_{l=0}^{n-1} (4l + 2) = 2n^2 \tag{18}$$

i.e., two electrons in the first period (n = 1), eight electrons in the second (n = 2), 18 electrons in the third (n = 3), and so on.

The first period is completed by only two elements (H, He), because the first level has a single sublevel that can be occupied by two electrons with opposite spin (because $n = 0$, l is also zero; m_l may vary from "l to $+l$ and is also zero; $m_s = +½$ and "½). The second period begins with an electron occupying the lowest energy orbital of sublevel $2s$ (Li) and terminates with the addition of the sixth electron of sublevel $2p$ (Ne). The pile-up procedure is analoguous for the third period, but in the fourth period this regularity is perturbed by the formation of transition elements (Sc-Zn) and by progressive pile-up of electrons in the $3d$ orbitals, whereas the fourth shell is temporarily occupied by one or two electrons. Pileup in the fourth level starts again with Ga, and the period terminates with the completion of the stable octet (Kr).

The same sequence is observed in the fifth period (Y-Cd transition elements). In the sixth period, the transition is observed over sublevel of the fourth shell, generating a group of elements (rare earths, La-Lu) of particular importance in geochemistry for the regularity of progression of reactive properties within the group (the most external shells are practically identical for all elements in the series; the progressive pile-up in the inner fourth shell induces shrinkage of the atomic radius, known as "lanthanidic contraction"). In the seventh period, the transition begins with actinium (Ac) and thorium (Th), whose external electrons occupy sublevel 6 d, and continues with the subsidiary series of protactinium (Pa-Lw), with electrons progressively occupying sublevel of the fifth shell.

Periodic Properties, Ionisation Potentials, and Energy of External Orbitals

In terms of energy, the existence of periodicity is easily understood by analysing the expressions that relate the energy of external orbitals to quantum number n.

Under the *hydrogen ion approximation* (ions of nuclear charge Ze with a single electron progressively occupying the various orbital levels), the potential acting on the electron by virtue of the nuclear charge is given by

$$\Phi_C = -\frac{Ze^2}{r}, \tag{19}$$

where Z is the atomic number and r is the distance from the nucleus expressed in Cartesian coordinates:

$$r = \left(x^2 + y^2 + z^2\right)^{1/2}. \tag{20}$$

Substitution of equation 1.19 into equation 1.9 leads to

$$\nabla^2 \psi + \frac{8\pi^2 m}{h^2}\left(E + \frac{Ze^2}{r}\right)\psi = 0. \tag{21}$$

The eigenvalues of equation 1.21 are

$$E_{(n)} = -\frac{2\pi^2 me^4 Z^2}{n^2 h^2}, \tag{22}$$

where $E_{(n)}$ is the energy of the various orbital levels in the hydrogen ion approximation.

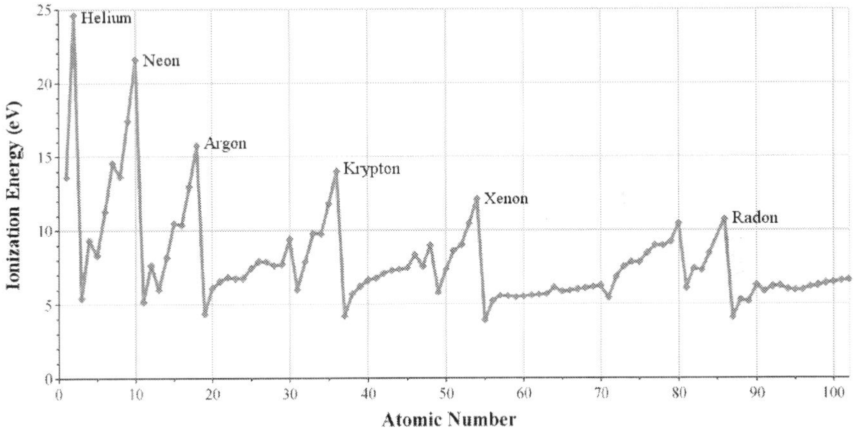

Figure: *First Ionisation potentials as functions of atomic number.*

The energy of external orbitals in real atoms (which is numerically equivalent to the first Ionisation potential expressed in eff) is reproduced by an empirical expression of the form:

$$E = \frac{(Z - S_e)^2 C_e}{n^2} = \frac{Z_{eff}^2 C_e}{n^2} \tag{23}$$

where S_e is a screen constant, depending on the number of inner electrons, and C_e is a conversion factor. Equation 23 establishes that, with an increase in the quantum number, we should observe a jump in the energy values of the external orbitals. We recall for this purpose that the external or "valence" electrons are those involved in the formation of atomic bonds (because the absolute values of their orbital energies are much lower than those pertaining to the inner electrons). The first Ionisation potentials plotted as functions of atomic number Z exhibit marked discontinuities, corresponding to the variations in the principal quantum number, and hence corresponding to the change of period in Mendeleev's periodic chart. Within the same period, we ob serve a progressive increase in the first Ionisation potential, with minor discontinuities for transition elements (due to the progressive filling of inner shells, which modifies the value of screen constant S_e in eq. 23). Recalling the definition of "valence electron" and considering that bonding energies are related to external orbitals, the periodicity of the reactive properties of the elements is readily understood in light of equations 22 and 23.

Generalities on the Concept of Chemical Bond in Solids

"There is a chemical bond between two atoms or groups of atoms in the case that the forces acting between them are such as to lead to the formation of an aggregate with sufficient stability to make it convenient for the chemist to consider it as an independent molecular species" (Pauling, 1948). We will consider four distinct types of bonds:

1. ionic bonds
2. covalent bonds
3. van der Waals bonds
4. metallic bonds

Although in nature it is rare to see solid compounds whose bonds are completely identifiable with one of the types listed above, it is customary to classify solids according to the dominant type of bond—i.e., *ionic, covalent, van der Waals,* and *metallic* solids.

In ionic solids, the reticular positions are occupied by ions (or anionic groups) of opposite charge. In NaCl, for instance, the Cl^- ions are surrounded by six Na^+ ions. The alternation of net charges gives rise to strong coulombic interactions.

In covalent solids, two bonded atoms share a pair of electrons. Covalent bonds are particularly stable, because the sharing of two electrons obeys the double duty of completing a stable electronic configuration for the two neighbouring atoms. In diamond, for instance, each carbon atom shares an electron pair with four surrounding carbon atoms, generating a stable giant molecule that has the dimension of the crystal.

In van der Waals solids, the reticular positions are occupied by inert atoms held together by van der Waals forces. The van der Waals bond is rather weak, and the bond does not survive a high thermal energy (*Fm3m* solid Ne, for instance, is stable below 20 K). The van der Waals bond also exists, however, in molecular compounds (e.g., solid CO_2). The reticular positions in this case are occupied by neutral molecules (CO_2). Within each molecule, the bond is mainly covalent (and interatomic distances are considerably shorter), but the various molecules are held together by van der Waals forces whose nature will be discussed later.

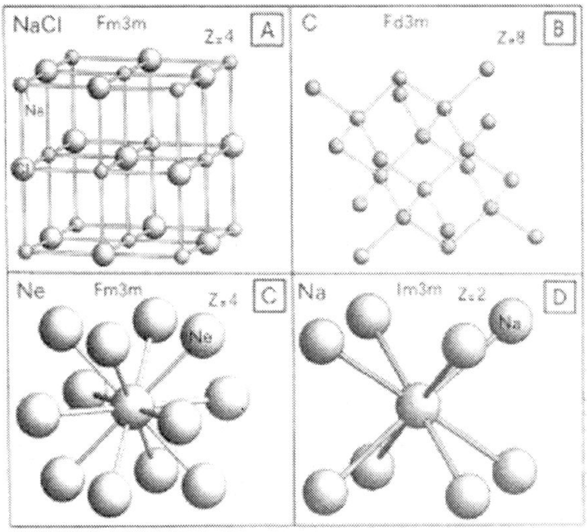

Figure: *Various types of solids. For each solid, a sketch of the bond is given.*

In metallic solids, the reticular positions are occupied by cations immersed in a cloud of delocalised valence electrons. In solid Na, for instance, the electron cloud or "electron gas" is composed of electrons of the *s* sublevel of the third shell. Note that the type of bond does not limit the crystal structure of the solid within particular systems. For example, all solids belong to the cubic system, and two of them (NaCl, Ne) belong to the same structural class (*Fm3m*).

Ionic Bond and Ionic Radius

The concept of "ionic radius" was derived from the observation of crystal structures understood as "regular arrays of ions of spherical undeformable symmetry." The structure of ionic crystals obeys pure geometrical rules and corresponds to the energetically optimal condition of an ion surrounded by the maximum number of ions of opposite charge.

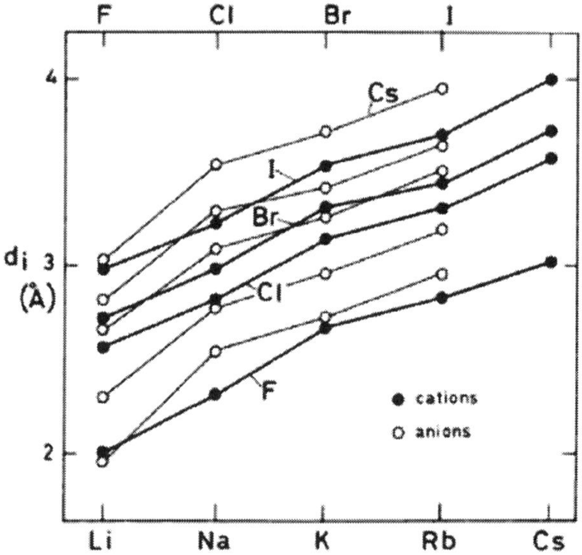

The assumption of "undeformability" makes it possible to consider interionic distance as a constant, at parity of ionic charge. The concept of ionic radius is thus related to that of "interionic distance" (d_i) by simple addition, according to

$$d_i = r_+ + r_-, \tag{24}$$

where r_+ and r_- are, respectively, the ionic radii of a cation and an anion at the interionic distance observed in a purely ionic crystal.

The concept that the ionic radius is relatively independent of the structure of the solid arose intuitively from experimental observations carried out on alkali halides, which are ionic solids par excellence. Significant parallelism within each of the two families of curves may be noted. This parallelism intuitively generates the concept of constancy of the ionic radius.

The first attempt to derive empirical ionic radii by geometrical rules applied to ionic compounds (and metals) was performed by Bragg (1920), who proposed a set of ionic radii with a mean internal precision of 0.006 Å. Further tabulations were proposed by Lande (1920), Wasastjerna (1923), and Goldschmidt et al. (1926).

However, we have already seen that the Schrödinger equation denies the existence of definite limits in the geometry of atomic orbitals, and that the presence of an electron in a particular portion of the space surrounding the nucleus is a probabilistic concept. These two facts apparently conflict with the concept of an ion as a spherical undeformable entity. But this conflict is only apparent, as may be deduced from the "electron density maps" obtained for crystalline compounds by X-ray diffraction techniques.

Concentric lines delineate zones of equal electron density (electrons per Å^3), increasing toward the nuclei (to a maximum of 70 electrons per Å^3 for Na^+ and 130 electrons per Å^3 for Cl^-). The absolute minimum electron density between the two nuclei defines the limits (hence the "dimensions") of the two ions. By integration of the electron densities, it can be calculated that 17.70 electrons are present in the region surrounding the nucleus of Cl^- and 10.05 electrons are present in the region surrounding the nucleus of Na^+.

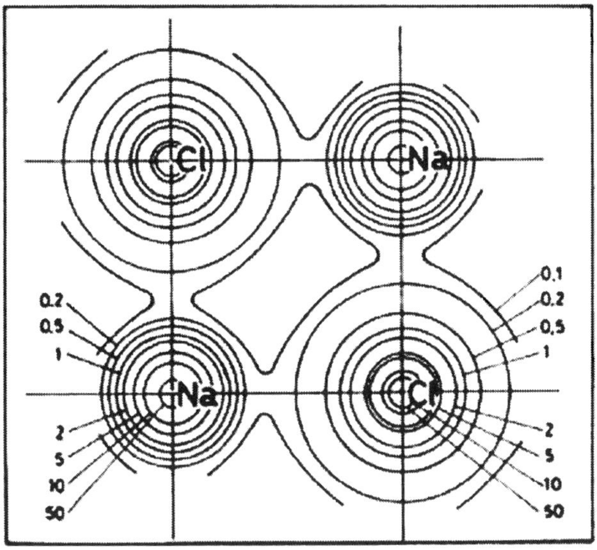

Figure: *Electron density (in electrons per Å^3) in the (100) plane of NaCl.*

These values, compared with the theoretical limits of 18 and 10 electrons, respectively, indicate that, within experimental approximation, there has been the complete transfer of one electron—and, hence, that a completely "ionic bond" is present in the cubic structure of NaCl.

We also observe that the electron density map defines almost perfect sphericity for both ions. The "ionic radii" of Na^+ and Cl^- in the crystalline cubic compound NaCl thus correspond to the radii (radial extensions) of the two elements in the condition of "ionic bond" when there is complete transfer of one electron from Na to Cl.

Pauling's Univalent Ionic Radii

The first attempt to derive ionic radii independently of the geometrical concept of simple additivity was made by Pauling (1927a). His method was based on the assumption that the dimension of an ion is determined by the distribution of its outer electrons. As we have already seen (eq. 23), the "effective nuclear charge" of an ion (Z_{eff})is lower than its atomic number, due to the shielding effect of inner electrons, expressed by screening constant S_e. Values of S_e may be obtained theoretically, and partially by results of X-ray diffraction studies. Based on such values, the univalent ionic radius (r_{+-}) can be expressed as

$$r_{+-} = \frac{C_n}{Z - S_e} = \frac{C_n}{Z_{eff}}, \tag{25}$$

where C_n is a constant determined by the principal quantum number of the outermost electrons in the inert gas configuration. Applying equation 25 to couples of isoelectronic ions such as NaF, KCl, RbBr, and CsI, which have, respectively, the same C_n, Pauling determined the dimension ratios of the isoelectronic ions in the corresponding solids and derived the corresponding ionic radii by combining obtained with observed values of interionic distance. For ions of the neon structure, for example, S_e has a value of 4.52, and the values of Z_{eff} for Na^+ and F^- are thus 6.48*e* and 4.48*e*, respectively. Because C_n is identical for the two ions, knowledge of the interionic Na^+–F^- distance (2.31 Å) gives the univalent radii solving the system

$$\frac{r_{F^-}}{r_{Na^+}} = \frac{Z_{eff,Na^+}}{Z_{eff,F^-}} = \frac{6.48}{4.48}, \tag{26}$$

$$r_{Na^+} + r_{F^-} = 2.31 \tag{27}$$

which gives 36 Å for F^- and 0.95 Å for Na^+. If we now continue to apply equation 1.25 to multivalent ions, assuming that C_n remains unchanged at the value of the intermediate inert gas, we obtain values for all ions with He, Ne, Ar, Kr, and Xe structures. These radii, known as "univalent ionic radii in VI-fold co-ordination," "do not have absolute values such that their sums are equal to equilibrium inter ionic distances" but rather "are the radii the multivalent ions would possess if they were to retain their electron distributions but enter into Coulomb interaction as if they were univalent". Univalent radii may be opportunely modified to obtain effective ionic radii in real structures:

$$r_Z = r_{+-} z^{-2/(n-1)}, \tag{28}$$

where *n* is the Born exponent and *z* is the ionic charge.

Covalent Bond and Covalent Radius

According to Lewis's (1916) definition, a "covalent bond" exists whenever two atoms are linked in a stable molecule by sharing two outer electrons. We distinguish between "homopolar" and "heteropolar" covalent bonds on the basis of whether or not the bonded atoms are of the same type. Pauling (1960) has shown that in covalent compounds the distance between two elements A and B, independent of the nature of the two atoms, is the same in all compounds containing both elements, regardless of the state of aggregation of the compound (i.e., solid, liquid, or gaseous).

Covalent radii are easily calculated by dividing by 2 the interatomic distances observed in homopolar covalent compounds (e.g., P_4, S_8, etc.). An internally consistent tabulation was derived by Sanderson (1960) from interatomic distances in heteropolar covalent gaseous molecules (e.g., $AlCl_3$, CH_4, SiF_4, etc.). The covalent radii of Sanderson are defined as "nonpolar," because they are obtained from compounds that do not have charge polarity.

Electronegativity and Fractional Ionic Character

The concept of "electronegativity" is widely used in geochemistry to define the combinational properties of the various elements in a qualitative manner. However, it is important to stress that this concept is largely empirical, because it varies from author to author and cannot be univocally quantified.

According to Pauling's (1932) definition, electronegativity is "the capacity of an atom of a molecule to attract the electrons of another atom." The electronegativity scale of Pauling (1960) is based on the differences in the binding energies in diatomic molecules AB with respect to homeopolar compounds AA and BB, according to

$$Z_A - Z_B = \gamma^{1/2}\left[E_{AB} - \frac{1}{2}(E_{AA} + E_{BB})\right]. \tag{29}$$

where χ_A and χ_B are the electronegativities of atoms A and B, E_{AB}, E_{AA}, and E_{BB} are the binding energies of compounds AB, AA, and BB, and operative constant γ is a term that constrains the adimensional χ_A and χ_B to change by 0.5 with each unit valence change in the first row of the periodic table ($\gamma^{1/2} = 0.208$).

According to Mulliken (1934, 1935), electronegativity is the algebraic mean of the first ionisation potential and of electron affinity.

According to Gordy (1946), electronegativity is represented by the value of the potential resulting from the effect of the nuclear charge of an unshielded atom on a valence electron located at a distance corresponding to the covalent radius of the atom. According to Sanderson (1960), electronegativity represents the mean electron density per unit volume of an inert element:

$$D_C = \frac{3Z}{4\pi_C^3}, \tag{30}$$

where Z is the atomic number and r_c is the covalent nonpolar radius.

Iczkowski and Margrave (1961) consider electronegativity as the local value of a potential defined by the partial derivative of the energy of the atom W with respect to ionic charge z:

$$Z_{(z)} = \frac{\partial W}{\partial z}. \tag{31}$$

The most immediate use of the concept of electronegativity is in the definition of "fractional ionic character" ($|_i$). The more or less ionic character of a bond is determined by the distribution of charges within the diatomic molecule. For a purely ionic bond, there is complete charge transfer between the two neighbouring atoms, whereas in a purely covalent bond there is no charge transfer. The relationship between electronegativity and percentage of ionicity in diatomic molecules has been proposed by several authors in different ways. According to Pauling (1932, 1960), the more or less ionic character of a bond is expressed by parametre $|_i$, derived as follows:

$$f_i = 1 - \exp\left[-\frac{1}{4}(Z_a - Z_B)^2\right]. \tag{32}$$

According to Gordy (1950), the same parametre can be obtained by applying

$$f_i = \frac{1}{2}(Z_A - Z_B). \tag{33}$$

Hinze et al. (1963) suggest that charge transfer in diatomic molecules obeys the principle of equality of energies involved in the process. If we adopt equation 31 as a definition of electronegativity and imagine that the amount of energy involved in the transfer of a fraction of charge dz from atom A to atom B is quantifiable as

$$dE_A = \left(\frac{\partial W_A}{\partial z}\right)_{ZA}^{dZ} \tag{34}$$

then, by application of the equality principle, we have

$$dE_A = \left(\frac{\partial W_A}{\partial z}\right)_{ZA} dz = \left(\frac{\partial W_B}{\partial z}\right)_{ZB} dz = dE_B. \quad (35)$$

A similar concept was developed by Sanderson (1960).

Electronegativity values according to Pauling and Sanderson are listed together with the first four Ionisation potentials and the electron affinities of the various elements.

At present, the most widely used definition of fractional ionic character in solid state chemistry is that of Phillips (1970), based on a spectroscopic approach. Phillips defined fractional ionic character as

$$f_i = \frac{E_i^2}{E_i^2 + E_c^2}, \quad (36)$$

where E_i ("ionic energy gap") is obtained from the "total energy gap" (E_g) between bonding and antibonding orbitals:

$$E_g = \frac{\hbar\omega_r}{(\varepsilon_- - 1)^{1/2}} \quad (37)$$

$$E_i = \left(E_g^2 - E_c^2\right)^{1/2} \quad (38)$$

where $\hbar\omega_p$ is the plasma frequency for valence electrons, ε_∞ is the optic dielectric constant, and E_c corresponds to E_g for the nonpolar system in the same row of the periodic table, with a correction for interatomic spacing.

Catlow and Stoneham (1983) have shown that ionic term E_i corresponds to the difference of the diagonal matrix elements of the Hamiltonian of an AB molecule in a simple LCAO approximation ($H_{AA} - H_{BB}$), whereas the covalent energy gap corresponds to the double of the off-diagonal term H_{AB}—i.e.,

$$E_g = \sqrt{(H_{AA} - H_{BB})^2 + 4H_{AB}}. \quad (39)$$

Ionic Polarizability and van der Waals Bond

The electron cloud of an ion subjected to an electric field undergoes deformations that may be translated into displacement "*d*"of the baricentres of negative charges from the positions held in the absence of external perturbation, which are normally coincident with the centres of nuclear charges (positive). The noncoincidence of the two centres causes a "dipole moment," determined by the product of the displaced charge (Z_e) and the displacement *d*. The displacement is also proportional to the intensity of the electrical field (F). The proportionality factor (α) is known as "ionic polarizability":

$$\mu_d = Z_e d = F\alpha. \tag{40}$$

Let us consider two neighbouring atoms A and B and imagine that the electronic charge distribution is, for an infinitesimal period of time, asymmetrical. In this period of time, atom A becomes a dipole and induces a distortion in the electron cloud of atom B by attracting electrons from its positive edge: atom B is then transformed, in turn, into a dipole. All the neighbouring atoms undergo the same process, generating attractive forces among themselves.

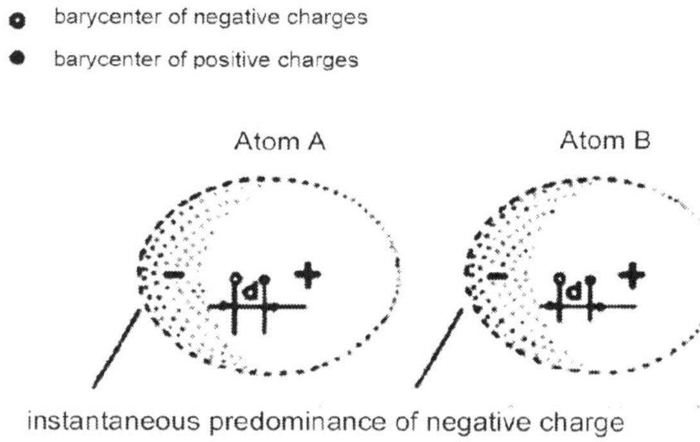

Figure: *Polarisation of charges in neighbouring atoms.*

Due to electron orbital motion, the induced forces fluctuate with a periodicity dictated by the rotational velocity of electrons. The *van der Waals bond* is the result of the induced attractions between fluctuating dipoles. The attractions diminish with increased molecular distance and, when distances are equal, increase with the number of electrons within the molecule.

Free-ion polarizability calculations have been proposed by various authors, based on the values of molecular refraction of incident light (R_m), according to

$$\alpha = \frac{3R_m}{4\pi N_0}, \tag{41}$$

where N_0 is Avogadro's number.

Free-ion polarizabilities, arranged in isoelectronic series, are listed. A semilogarithmic plot of the listed values reveals a functional dependence on atomic number. This dependence, quite marked, allows us to estimate the free-ion polarizability for ions for which there are no precise experimental data.

Crystal Radius and Effective Distribution Radius (EDR)

The first satisfactory definition of crystal radius was given by Tosi (1964): "In an ideal ionic crystal where every valence electron is supposed to remain localised on its parent ion, to each ion it can be associated a limit at which the wave function vanishes. The radial extension of the ion along the connection with its first neighbour can be considered as a measure of its dimension in the crystal (crystal radius)." This concept is clearly displayed in which the radial electron density distribution curves are shown for Na^+ and Cl^- ions in NaCl. The nucleus of Cl^- is located at the origin on the abscissa axis and the nucleus of Na^+ is positioned at the interionic distance experimentally observed for neighbouring ions in NaCl. The superimposed radial density functions define an electron density minimum that limits the dimensions or "crystal radii" of the two ions. We also note that the radial distribution functions for the two ions in the crystal (continuous lines) are not identical to the radial distribution functions for the free ions (dashed lines).

The "crystal radius" thus has local validity in reference to a given crystal structure. This fact gives rise to a certain amount of confusion in current nomenclature, and what it is commonly referred to as "crystal radius" in the various tabulations is in fact a mean value, independent of the type of structure.

The "crystal radius" in the sense of Tosi (1964) is commonly defined as "effective distribution radius" (EDR). The example given shows radial electron density distribution curves for Mg, Ni, Co, Fe, and Mn on the M1 site in olivine (orthorhombic orthosilicate) and the corresponding EDR radii located by Fujino et al. (1981) on the electron density minima. It also must be stressed that EDR may assume different values along various crystallographic directions. This is clearly evident if we compare electron density maps obtained by calculation of electron density functions of neutral, spherically averaged atoms (known as Independent Atom Model or IAM) with observed total electron densities. For instance, makes this comparison for BeO: subtraction of IAM densities (B) from total electron densities (A) results in a deformation density chart (C), which shows how the electron density of the IAM reference model has migrated due to bonding effects. (Solid contours in the figure are positive accumulations of charges; short dashes are negative; long dashes mark unperturbed zones.)

Chapter 2

The Constitution of Liquids and Surface Tension Phenomena

The Epoch-Making Work of LAUE and BRAGG has given us a clear insight into the structure of crystalline bodies. In the case of inorganic crystalline substances, it has been found that the atoms are arranged according to definite lattices so that certain groups of atoms are repeated in a regular manner. In this arrangement, however, there is nothing to indicate where one group ends and the next begins. Thus, in a crystal of common salt, the chlorine and sodium atoms alternate in regular order, but are not grouped together into pairs. Any division of the crystal into molecules is thus purely arbitrary. We must therefore conclude that solid bodies are not built up of molecules, but of atoms arranged in definite ways; or rather, we should regard a crystal as a single large molecule.

From this viewpoint the forces holding solid bodies together (cohesion, etc.) are chemical forces, and phenomena such as evaporation are strictly chemical phenomena. So far, in the study of crystal structure, only inorganic bodies have been investigated. It is certain that in solid organic bodies the atoms are arranged in groups which are usually identical with the molecules found when the substances exist as vapours. We may call these *group molecules*. There are good reasons for believing that in crystals of an organic substance these group molecules are held together by chemical forces (secondary valence) to form single large molecules.

A similar theory is now proposed for the structure of liquids. Each atom in a liquid is regarded as being chemically combined with all adjacent atoms. This chemical union may in general be of two kinds, characterised by *primary* and *secondary* valences. The atoms held together by primary valence usually constitute *group molecules*, in which the atoms are more or less rigidly attached to each other in definite arrangements. The secondary valence serves to hold the group molecules together. In very many cases (such as most inorganic liquids) the group molecules are very much simpler than the usual "chemical molecules" and may often be entirely absent. Thus, in such substances as water, concentrated sulphuric acid or in molten salts or metals, there are probably no group molecules at all; the atoms being held together by secondary valences.

A drop of liquid is thus regarded as a single large molecule. In fact, the whole earth, including the oceans (but excluding the atmosphere) may be looked upon as a single molecule. The most essential difference between liquids and solids is the mobility of the former. The chemist is already familiar with a mobility between different parts of a molecule and has given the name tautomerism to this phenomenon. The mobility of liquids is therefore a result of tautomeric re-arrangements of the atoms. Evaporation, condensation, freezing, melting, solution, adsorption, and surface tension, and even viscosity, are thus chemical phenomena, and chemical knowledge already available may be directly applied in their study.

This theory has been applied with marked success in the study of the adsorption of gases by solids. Adsorbed gases are regarded as being chemically combined (either primary or secondary valence) with the *atoms* forming the surface of the solid. The adsorbed film thus usually consists of a *single* layer of atoms or group molecules, which forms a continuation of the structure of the solid. The effect of such films on the catalytic activity of surfaces (as, for example, platinum) depends entirely upon the nature and the arrangement of the atoms forming the actual surface of the solid.

In a similar way a theory of surface tension is now proposed in which the structure of the *surface layer of atoms* is regarded as the principal factor in determining the surface tension (or rather surface energy) of liquids. This theory is supported in the most remarkable way by all available published data on the surface tension of organic liquids.

According to this theory, the group molecules of organic liquids arrange themselves in the surface layer in such a way that their active portions are drawn inwards, leaving the least active portion of the molecule to form the surface layer. By "active portion" of a molecule is meant a portion which is characterised by a strong stray field (residual valence). Chemical action may be assumed to be due to the presence of electromagnetic fields surrounding atoms. Surface tension (or surface energy) is thus a measure of the potential energy of the electromagnetic stray field which extends out from the surface layer of atoms. The molecules in the surface layer of the liquid arrange themselves so that this stray field is a minimum.

The surface energy of a liquid is thus not a property of the group molecules, but depends only on the *least active portions of the molecules* and on the manner in which these are able to arrange themselves in the surface layer. In liquid hydrocarbons of the paraffin series, the molecules arrange themselves so that the methyl groups (CH_3) at the ends of the hydrocarbon chains form the surface layer. The surface layer is thus the same, no matter how long the hydrocarbon chain may be. As a matter of fact, the surface energy of all these many different substances from hexaine to molten paraffin, have substantially the same surface energy—namely, 46 to 48 ergs per square centimetre, although the molecular weights differ very greatly.

If, now, we consider the alcohols such as CH_2OH, C_2H_2OH, etc., we find that their *surface energies are practically identical with those of the hydrocarbons.* The reason for this is that the surface layer in both cases consists of CH_3 groups.

With such substances as CH_3NO_2, CH_3I, we find that the surface energy is much greater than that of the hydrocarbons. This is due to the fact that the volume of the I or the NO_2 is so great that the surface cannot be completely covered by the CH_3 radicals. The forcing apart of these groups increases the surface energy.

Particularly interesting relations are found with benzol derivatives. In benzol itself, the group molecules arrange themselves so that the benzol rings lie flat on the surface, since the flat sides of these rings are the less active portions of the molecules. The surface energy of benzol is about 65 ergs per square centimetre.

If, now, an active group, such as OH, is substituted for one of the hydrogens in the benzol (forming phenol or carbolic acid), this group is drawn into the body of the liquid, *tilting the bensol ring up on edge* and raising the surface energy to about 75 ergs per square centimetre, which corresponds to the activity of the perimetre of the benzol ring. Thus *any active group* strong enough to tilt the ring up on edge raises the surface energy to about 75. Two active groups side by side (ortho position) have no greater effect than one. But two active groups opposite one another (para position) cannot both go wholly below the surface, so that the surface energy then becomes abnormally large (about 85 in case of para nitrophenol). The substitution of methyl or ethyl groups in the benzol ring lowers the surface energy except where an active group in an adjacent position draws these groups below the surface.

Some of the best evidence in support of the new theory is derived from experiments on thin films of oil on water or mercury. Oleic acid on water forms a film one molecule deep, in which the hydrocarbon chains stand vertically on the water surface with the COOH groups in contact with the water.

Acetic acid is readily soluble in water because the COOH group has a strong secondary valence by which it combines with water. Oleic acid is not soluble because the affinity of the hydrocarbon chains for water is less than their affinity for each other. When oleic acid is placed on water the acid spreads upon the water because by so doing the COOH *can dissolve in the water* without separating the hydrocarbon chains from each other.

When the surface on which the acid spreads is sufficiently large, the double bond in the hydrocarbon chain is also drawn down on to the water surface, so that the area occupied is much greater than in the case of the saturated fatty acids. Oils which do not contain active groups, as for example pure paraffin oil, do not spread upon the surface of water.

The measurement of the area of water or mercury which can be completely covered by a given amount of a substance affords an accurate method of determining the shapes of group molecules. Thus it is found that the molecules of stearic acid on a surface of water have a length of about 23×10^{-8} cm. and cover an area of 24×10^{-16} sq. cm. These measurements prove that the molecules are not spherical but are much elongated.

This theory also affords an explanation of the mechanism by which colloids are formed. If a film of closely packed oleic acid molecules covers the surface of water to which sodium hydroxide has been added, OH groups are adsorbed by the COOH radicals, causing an expansion of the lower side of the film without a corresponding expansion of the upper side. This results in the bulging of the film downwards in spots so that it finally detaches itself in the form of particles, the outer surfaces of which consist of COOH groups together with the adsorbed OH, while the interior consists of the long hydrocarbon chains.

The size of the colloidal particles is determined by the difference in size between the two ends of the molecules just as the size of an arch is dependent upon the relative sizes of the two ends of the stones of which the arch is constructed. In the final publication of this paper the bearing of the shapes of molecules on the formation of colloids will be discussed in detail.

Forces Near the Surfaces of Molecules

In the early development of the kinetic theory, during the last century, it was usual to consider the molecules of gases as rigid spheres, which exerted no forces on one another except at the instant of contact. In this way it was possible to account quantitatively with fair accuracy not only for the effects of pressure and temperature on the volume, but also for the viscosity, heat conductivity, and diffusion of gases, and other phenomena involving the free paths of the molecules.

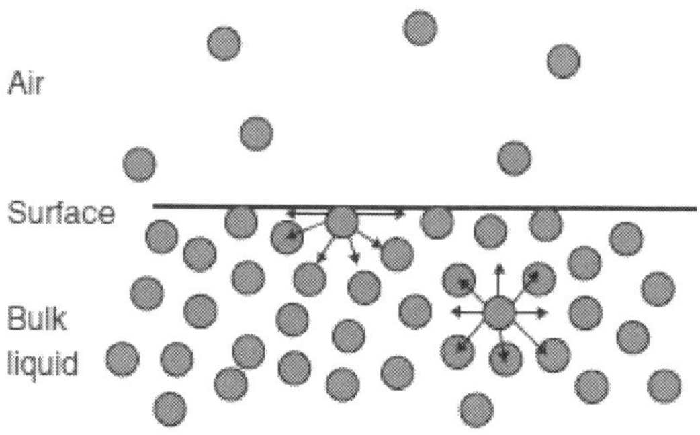

By considering also attractive forces which varied with some power of the distance between the centres of the molecules, van der Waals was able to account for the main features of the transition from the gaseous to the liquid state, and also explained in a satisfactory manner the departures from the ideal gas laws which are shown by all gases at high pressures.

As new properties of matter are brought into consideration, or as a deeper understanding of them is needed, it is necessary to postulate additional properties of the molecules, and to know more accurately the nature of the forces acting between them. For example, to deal with the properties of electrolytes it is necessary to consider electric charges on the molecules or ions. In 1912 Debye showed that the electrical properties of many dielectrics could be accounted for by assuming that the centre of gravity of the electrons in a molecule did not coincide with the centre of gravity of the positive charges of the nuclei. Thus the molecule possessed what we call a dipole moment, which can be measured as the product of an electric charge by the distance through which it is displaced.

It is found, for example, that a molecule of water possesses a dipole moment of 1.8×10^{-18} electrostatic units. A moment of this magnitude could be produced by having an electron at a distance of 0.37×10^{-8} cm. from a proton. Since there are 10 electrons in a water molecule, it is only necessary that the centre of gravity of these electrons shall lie at a distance of 0.037×10^{-8} cm. from the centre of gravity of the three nuclei in the molecule. Thus displacements of the electrons to distances minute compared with the diametre of the molecule can account for even the largest values of dipole moments that have been observed.

H. B. Hardy in 1912 and the writer in 1916 pointed out that the fields of force around many chemical molecules must be very unsymmetrical, and that this resulted in an orientation of the molecules at the surfaces of liquids and in adsorbed films on solids. It was also found that the shapes which the molecules could assume, and which were determined by their chemical structures, were of great importance in connection with such properties as surface tension.

Within recent years, particularly through the work of Debye and his students, a great deal of knowledge has been accumulated regarding the electrical forces in liquids and solutions which contain ions and dipoles Particularly in the study of electrolytes has this work cleared up many of the most serious difficulties of earlier theories.

There is now no need of drawing distinctions between chemical and physical forces. The chemist recognises that many of the forces with which he has been dealing are electrical in nature, for, after all, molecules are built up of electrons and positive nuclei, and their interactions must necessarily be electrical. However, the simple classical theory of electrical forces based on Coulomb's law is wholly insufficient to account for chemical properties. With charged particles obeying Coulomb's law only, the minimum of potential energy is obtained when the positive and negative particles coincide. There must be something corresponding to repulsive forces holding the particles apart.

Repulsive Forces between Molecules

In many of the later developments of the kinetic theory where the molecules were regarded as rigid spheres, the repulsive forces which must counteract the attractive forces were assumed to act only during collisions-that is, at the instant of contact of the spheres. No explanation was attempted of the internal forces within the molecule which could result in this repulsion between molecules.

In order to account for the compressibility of solids, Born and Landé have assumed that the repulsive force between molecules (or ions) varies inversely as a power of the distance between the centres of the molecules. Thus in the case of halide salts of the alkalies, such as NaCl, KI, etc., they showed that the compressibility indicated that the repulsive force varied inversely as the tenth power of the distance, while the attractive force resulting from the charges on the ions varied inversely as the square, in accord with Coulomb's law.

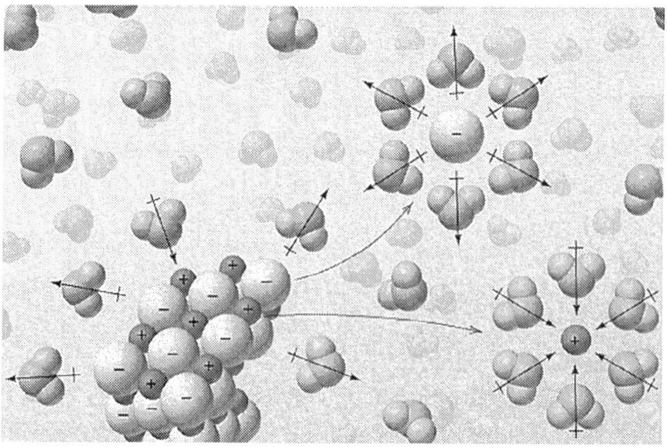

Born also attempted to show that the inverse tenth power law for repulsion could be deduced by assuming that the electrons and the ions were arranged with cubic symmetry as at the corners of a cube. This attempt, however, must be regarded as a failure for two reasons. First, the method of mathematical expansion which was used is not legitimate when applied to distances as short as those between ions in salts; and, second, Born assumed that the cubical ions were all orientated so that their edges were parallel. This kind of orientation would be unstable under the forces assumed, for these forces would tend to orientate one cubical ion so that its corner would come opposite the face of another ion, and in this case the repulsive force would be replaced by an attractive force which would act in addition to the force between the ionic charges. Although such orientation of ions as Born assumes might conceivably exist in solid crystals from causes which he forces not discuss, it is clear that it cannot be the cause of the repulsive forces, since the elimination of the orientation by melting the salt does not greatly alter the density or the compressibility.

Debye has developed a theory which seems to give a better picture of the repulsive forces. Assuming the electrons to move in orbits in accord with Bohr's theory, he finds that the electric field around a molecule is a pulsating or partly oscillating field increasing in amplitude very rapidly as the surface of the molecule is approached. Thus when an electron approaches a molecule it acquires, in addition to any motion of translation it may have, an oscillatory motion due to the pulsating field. Debye shows that this oscillatory motion tends to be 180° out of phase with the field producing it, and then shows, that because of the non-uniformity of the electric field near an atom the resulting force will be one of repulsion.

On this theory the repulsive forces between molecules are due to the perturbations in the electron orbits in one molecule, caused by the motions of the electrons in their orbits in the neighbouring molecules. The repulsive force thus does not originate from a molecule as a whole, but comes from the close approach of electrons in the two molecules. This would seem to indicate that we may profitably regard these forces as *surface forces*, and should express them as functions of the distances between the surfaces of the molecules rather than in terms of the distances between the centres. This conception affords a justification, to a large extent, of the theory which regarded molecules as rigid elastic spheres. Even if we accept Debye's picture of the origin of repulsive forces, it really only refers the cause back to the quantum theory, by which the stable orbits of the electrons in the molecule are determined. At present, however, problems of this kind are treated by wave mechanics instead of by a consideration of orbits of electrons. For a few simple cases, such as interactions between hydrogen molecules, Hund and London have been able to calculate the forces by means of wave mechanics. But it will probably be a long time before definite knowledge of repulsive forces can be obtained in this way for more complicated molecules.

The quantum theory teaches that each electron in an atom is characterised by four quantum numbers, which are integers. Not all quantum numbers are possible; there are, for example, certain conditions (selection rules) which they must satisfy. The normal atom or molecule, with which the chemist is concerned, is one in which the electrons are so arranged that the total energy is a minimum, subject to the quantum rules that must be fulfilled.

It seems that the factor of greatest importance to the chemist is that which is known as the Pauli principle, which states that no two electrons in an atom can have the same combination of four quantum numbers. It is this principle which underlies the Periodic Table of the elements, and in a similar manner determines the arrangement of electrons in a molecule. It is through the development of methods of this kind that our greatest hope lies in obtaining a better understanding of the nature of repulsive forces.

For most practical purposes it is best for the chemist to regard molecules of organic substances as having shapes suggested by chemical structural formulas, and as having surfaces which are rather rigid so that strong repulsive forces are called into play for only moderate deformations. For many purposes it is still useful to look upon the surfaces as completely rigid.

Attractive Forces between Molecules

Whereas the quantum theory seems essential in an understanding of the repulsive forces, the classical laws, such as the Coulomb law, are largely sufficient for dealing with the attractive forces between molecules. For example, the Debye-Hückel theory of electrolytes is based on the Coulomb law, the Poisson equation, and the Boltzmann equation—all laws of classical mechanics. In discussing forces between molecules, Debye has classified molecules in three groups: "(1) ionic, (2) polar and (3) non-polar."

The ionic type is represented by electrolytic ions, such as K^+, NO_3^-, etc., and gaseous ions. In these molecules the number of electrons is unequal to the number of positive charges of the nuclei of atoms, so that the molecule as a whole has an electric charge which is some multiple of e, the charge of an electron. The polar type is represented by molecules which have a dipole moment-that is, uncharged molecules in which the centre of gravity of the negative charges does not coincide with that of the positive. Non-polar molecules thus include all molecules which are uncharged and in which the centres of gravity of the positive and negative particles coincide. From Debye's point of view such molecules may correspond to quadrupoles, octopoles, etc. For example, positive and negative charges arranged alternatively at the corners of a square will give a quadrupole, at the corners of a cube an octopole. In forming clear conceptions of the attractive forces between molecules it is useful to consider the actual magnitude of the forces involved.

Ions

The electric field at a distance r from a univalent ion having a charge e is $e/år^2$, where å is the dielectric constant of the medium. The charge of the electron, e, is equivalent to 1.43×10^{-7} volt centimetre and thus the electric field is $1.43 \times 10^{-7}/år^2$ volts per centimetre, and the electric potential corresponding to this force is 1.43 × 10-7/år. If we take for r the value 3×10^{-8} cm., which corresponds roughly to the distance between two ions in contact, we find the electric field to be 1.6×10^8 volts per centimetre and the potential to be 4.8 volts. Taking for å the value 80, corresponding to the dielectric constant of water, we would obtain values only one eightieth of those given. However, it is doubtful if the conception of dielectric constant is entirely justified at distances as short as this for we can hardly speak of two molecules in contact as being in a medium having known electrical properties. At distances several times greater than the molecular diametre the use of dielectric constant in this way is justified. The applications of the Debye-Hückel theory to electrolytes prove that the forces at these larger distances are of vital importance in an understanding of the properties of ions. According to the Coulomb law the force between two ions varies inversely as the square of the distance. In electrolytic solutions the increased concentration of positive ions around negative ions, and vice versa, causes the force between two given ions to decrease with increasing distance more rapidly than according to the inverse square law.

Polar Molecules

At a distance r from a dipole molecule having an electric moment μ the electric potential is $\mu \cos \theta/r^2$, where θ is the angle between the radius vector r and the axis of the dipole. It can thus be found that at a distance r along the dipole axis the electric field is $-2\mu/r^3$, this force being in a direction of the polar axis. At a distance r in the equatorial plane of the dipole the electric force is also in a direction parallel to the polar axis but has the magnitude $+\mu/r^3$.

Debye (9) has shown that in general because of thermal agitation a dipole (or quadrupole) molecule in a gas or liquid is changing its orientation so fast that we must not regard the force produced by a dipole as a steady force, but a rapidly fluctuating force, and under these conditions many of the effects produced by the force are proportional to E^2 or the mean square field. The effective force can thus vary inversely as the sixth power of the distance from the dipole.

Principle of Independent Surface Action

From the foregoing analysis of the forces acting between molecules of various types, we conclude that there is a large class of substances for which we are justified in regarding the forces between two molecules in contact as being dependent mainly on the nature of the surfaces of the molecules which are in contact. This principle of independent surface action will always be only an approximation to the truth, but there are many cases where it applies with sufficient accuracy and so greatly simplifies the problems of the interactions between molecules that useful results are obtained in problems so complex that no solution is otherwise possible. Let us consider some of these problems.

The theory of adsorption in monomolecular films on solids and liquids is an example of the application of this principle. According to this theory, the force which holds an adsorbed molecule or atom on a surface depends on the character of the surface of contact between the molecule and the solid. If a second layer of molecules should form, the forces holding the molecule to the second layer are thus entirely different from those holding the molecules of the first layer. The rates of evaporation of the molecules from the first layer or from the second layer will differ greatly, especially since they depend upon the magnitude of the forces according to an exponential relation of the Boltzmann type.

Two general cases must be considered. If the forces holding the molecules in the first layer are greater than those holding those of the second, there will be a wide range of pressures of the adsorbed gas for which the film will never exceed one molecule in thickness. We thus arrive at the conception that a certain fraction, θ, of the surface is covered by adsorbed molecules, and that the properties of this adsorbed film depend primarily on θ. If the forces between these molecules and the underlying surface are very large compared to those that act between adjacent molecules, then, for example, we may conclude that the rate of evaporation of the adsorbed molecules will be proportional to θ. This simple conception leads to an adsorption isotherm which has been found to hold experimentally in a large number of cases. However, if the forces that act between adjacent adsorbed molecules are not negligible, there will be large deviations from this simple law, but the theory may be easily extended to take into account such forces. The forces may be those of attraction, or, if the molecules become dipoles as a result of the adsorption, or if they become crowded, their interaction may cause re-pulsive forces.

The second important case to be considered is that in which the forces that hold the second layer of molecules are greater than those which hold the first; or, more generally, the case in which the forces acting between adjacent adsorbed molecules are greater than those with which each of these is held on the underlying surface. There is then great difficulty in getting any appreciable number of molecules in the first adsorbed layer, and single molecules or atoms or groups of molecules in the first layer act as nuclei from which large aggregates or crystals may develop. The condensation of cadmium or mercury vapour on cooled glass surfaces is an example of this type. It is possible to formulate the problem quantitatively and to express the number of nuclei that will form per second as a function of temperature and pressure. The calculated values are in good agreement with those obtained by experiment. If definite numbers of isolated copper atoms are evaporated on to a clean surface, each one serves as a nucleus for the formation of cadmium crystals, when the temperature of the surface and the pressure of the cadmium vapour are carefully regulated, and thus by dark field illumination of the surface the copper atoms can be directly counted. Some experiments made several years ago by Mr. Harold Mott-Smith have shown that this method can be developed to count atoms just as the C. T. R. Wilson method can be used to count ions.

In case the adsorbed atoms on a metallic surface are electrically charged or acquire large dipole moments, electrical forces are brought into play which may make the principle of independent action inapplicable. For example, the rate of evaporation of electrons from a tungsten surface at high temperature is enormously increased by the presence of adsorbed thorium or caesium atoms on the surface. But the increase in the number of electrons which evaporate is not even approximately proportional to the amount of thorium or caesium which is present on the surface. The same is true for the evaporation of positive caesium ions from a tungsten surface. In both of these cases, the heat of evaporation of electrons or ions varies approximately linearly with θ, the fraction of the surface covered by the adsorbed atoms, and thus, according to the Boltzmann equation, the logarithm of the rate of evaporation of electrons or ions varies linearly with θ. In the case of the evaporation of neutral atoms-for example, atoms of oxygen from adsorbed films on heated tungsten filaments—the rate of evaporation is much more nearly proportional to θ.

In studies of the surface tensions of organic liquids and of the properties of adsorbed films of organic substances on water and of oil films on water, the principle of independent action has proved itself particularly useful. It gives immediately a simple reason for believing that these films should rarely be more than one molecule in thickness; or rather, that if they are more than one molecule in thickness, their properties will be markedly different from those of films in which the surface is not completely covered with a single layer. The theory also gives immediately reasons for believing that adsorbed films of many organic substances on water will consist of orientated molecules. Thus the area on a water surface occupied by molecules of the various fatty acids is dependent on an interaction between the carboxyl group and the water, and is independent of the length of the hydrocarbon chain. This simple result, proved by experiment, demonstrates that the molecules are orientated so that the carboxyl group is in contact with the water, leaving the hydrocarbon tails to form a layer above the carboxyl group which will have the properties characteristic of a liquid hydrocarbon.

The theory is equally applicable in pure organic liquids. The total surface energies (surface tension extrapolated to the absolute zero) of hexane and of hexyl alcohol are practically identical. This is readily explained by the orientation of the molecules which prevents the hydroxyl groups from coming in contact with the surface, so that in both cases the actual surface is that of a pure hydrocarbon. Studies of the surface tension of aqueous solutions of various aliphatic compounds in water prove that in concentrated solutions the surface becomes covered with a closely packed monomolecular film consisting of orientated molecules like those in oil films. With sufficiently dilute solutions, however, the amount of substance adsorbed in the surface is not enough to cover the whole surface with closely packed molecules. Under these conditions, the hydrocarbon chain lies flat in the water surface and the energy needed to transfer any molecule from the surface to the interior increases by a definite amount for each additional CH_2 group in the hydrocarbon chain. This is an excellent example of the application of the principle of independent surface action, each CH_2 group producing its effect independently of the others.

Similar views prove useful in studies of many other properties of matter. Consider, for example, jellies made by amounts of gelatin or soaps in concentrations less than 1 per cent by weight. The elasticity proves the existence of a continuous frame-work of molecules in contact, extending throughout the liquid. The time of relaxation of such jellies serves as a measure of the rate at which the molecules which form the chains separate or evaporate from one another. Guided by these views, I made some experiments, over ten years ago, to determine the diametres of the cross sections of the fibres or rods that must be the elements of the rigid frame-work.

For this purpose I made up some dilute gelatin jellies on filter paper and measured the rate at which water could be forced through the jellies by applying a definite pressure. By a modification of Stokes' law which gives the rate of fall of small spheres in liquids, it was possible to calculate the force necessary to move small cylinders of various diametres through a liquid. This law was then checked experimentally by measuring the rate at which water passed through a column of glass-wool having fibres of known size. Applying this law to the case of the motion of water through a gelatin jelly, it was thus possible to calculate the size of the fibres. Only rough experiments were made, but the results showed clearly that the diametre of the fibres was approximately 10^{-7} cm.

I believe that there is real justification for dealing with molecules of this character, at least as a first approximation, as though they followed the same laws as bodies of large size, such as the fibres of glass-wool. Einstein showed, years ago, that Stokes' law could be used, approximately, for the study of the rate of migration of ions through water solutions. I believe that a thorough quantitative study of the forces necessary to drive water through various jellies should give much valuable information as to the structure of these jellies. Very useful pictures of the mechanism involved in the viscosity of liquids and of diffusion in liquids and solids may be had by considering that molecules in contact exist in two states: one in which the surfaces are rigidly connected, the other in which they are entirely free to move.

The behaviour of the molecules is thus analogous to that of a gas condensing on a solid; the molecules of the gas strike the surface, remain adsorbed for a certain time, and then evaporate off again. Applying this conception to molecules in liquids, the motion of the molecules past each other, involving viscosity and diffusion, depends upon the relative times during which the molecules are in this rigid, or in this mobile contact. These times can be calculated from an equation of the Boltzmann type in terms of the energy difference between the two states. This theory accounts for the frequent occurrence of temperature coefficients of viscosity and diffusion, which agree with the Boltzmann equation. It seems also to account for the fact that the viscosity of different members of a series of hydrocarbons increases in geometrical proportion to the length of the chain. It would seem that the principle of independent surface action should afford a means for the development of a simple theory of viscosity which would take fully into account the shapes and sizes of the molecules and of the different chemical groups contained in the molecules.

The principle has also found many applications in connection with the mechanism of heterogeneous chemical reactions. The interaction of oxygen and hydrogen and of carbon monoxide and oxygen at low pressures in contact with a heated platinum filament has shown that equations which are based directly on this principle are in excellent agreement with the experimental results over very wide ranges of pressure and temperature. For example, the experiments showed that at low temperatures the reaction velocity was accurately proportional to the partial pressure of oxygen and inversely proportional to the pressure of carbon monoxide. This behaviour could be completely accounted for by assuming that the rate of reaction was determined by the rate at which oxygen molecules could reach holes in the films of adsorbed carbon monoxide which were left by evaporation of carbon monoxide molecules, or which were formed by the removal of the carbon monoxide by the oxygen which reached these holes. The poisoning effect of the carbon monoxide was due to the fact that the opportunity for the oxygen to reach the holes was decreased if the carbon monoxide molecules were able to fill up the holes before the oxygen arrived. It seems difficult to reach quantitative agreement with these experiments except by the application of the principle of independent surface action.

The interaction of hydrogen and oxygen in contact with tungsten filaments at temperatures ranging from 1500° to 2500°K is another illustration of a similar kind. Oxygen acts to form WO_3 at a rate proportional to the oxygen pressure. Oxygen atoms *in the adsorbed oxygen film* do not react with one another and with the tungsten to form WO_3 even at the highest temperatures. At 1500°K the life of oxygen atoms on the surface is of the order of years; at 1860°K it is about 25 minutes; and at 2070°K it is 15 seconds. When the atoms leave the films at the higher temperatures they do so as free atoms, not as molecules. Thus, even in the presence of minute pressures of oxygen, the tungsten surface is practically completely covered with a single layer of oxygen atoms. Oxygen molecules which strike this surface condense on it, but evaporate from this second layer at a relatively high rate; but while thus adsorbed, they move freely over the surface and are able to fill up any holes that form in the first layer. Also the molecules, while adsorbed in the second layer, have a certain probability of interacting with those of the first layer and with the underlying tungsten to form WO_3, and the holes thus formed by the removal of oxygen from the first layer are soon filled up by the migration of the molecules of the second layer.

Hydrogen molecules are not able at any temperature to react directly with the oxygen in either of the first or second layers, but if they reach the tungsten at one of the holes in the first layer, they react immediately with the adjacent oxygen. Thus when the oxygen pressure falls to a certain low critical value, the hydrogen completely removes all the oxygen from the surface, and it does this suddenly. The hydrogen which then strikes the film is dissociated into atoms to an extent that depends upon the temperature, and the atomic hydrogen formed goes to the bulb and reacts with the WO_3 which has previously been deposited there.

This theory which can readily be stated quantitatively appears to be in complete agreement with all the experimental facts, so that again we find support for the principle of independent surface action. The molecules in adsorbed films on solutions of organic substances in water are frequently in a state of a two-dimensional gas. Oil films on water may exist as solid or liquid films. In the case of so-called expanded films, the heads of the molecules act as a two-dimensional gas, while the tails form a two-dimensional liquid. The molecules of oils adsorbed on solids, such as those that are responsible for some of the lubricating properties of oils, and those involved in the phenomena of flotation of ores, usually show little or no tendency to move over the surface. In other words, the molecules appear to be attached rigidly to the surface.

Many of the equations that were developed in the quantitative studies of the velocities of heterogeneous reactions assumed that bare spots on the surfaces were distributed over the surface according to statistical laws. Such statistical distribution would not occur unless the molecules possessed a certain degree of mobility over the surface, for the molecules that are removed by the reaction are those that are adjacent to holes already existing. Volmer and others have shown experimentally that adsorbed atoms frequently possess great mobility even on solid surfaces and thus act like two-dimensional gases of high viscosity.

I believe that the principle of independent surface action will be useful in studying many properties of organic substances which have hitherto been too complicated to be treated quantitatively. Within recent years I have made some attempts of this kind. When a liquid is separated into two parts, along a surface having an area of 1 sq. cm., two new surfaces with a total area of 2 sq. cm. are formed. The surface tension or free surface energy measures the work done per unit area in forming the new surfaces. The total surface energy γ, which is equal to the free surface energy extrapolated back to the absolute zero, represents the total energy change per unit area. For all the pure hydrocarbons, such as pentane and nonane, γ is equal to about 48 ergs $cm.^{-2}$ The fact that this value is practically independent of the length of the hydrocarbon chain proves that the surface forces are very nearly uniform over the whole of the hydrocarbon molecule.

When a hydrocarbon, such as hexane, evaporates, molecules pass from the interior of the liquid into the vapour. If we consider a drop of hexane liquid to be removed from a large volume of the liquid into the free space above it, the work that would be needed to form this drop would be equal to $S\gamma$ where S is the surface area of the drop. Since we are regarding molecules as having surfaces possessing certain properties, we may say that the work necessary to remove a single molecule of hexane from the liquid, which is the latent heat of evaporation per molecule, will also be equal to $S\gamma$, where S is now the surface area of the molecule and γ is the surface energy of the molecule per unit area. The molecular surface S for the molecule of vapour may be calculated from the molecular volumes (molecular weight divided by density), assuming the surfaces to be the same as in the case of closely packed spheres. This assumption would probably be quite accurate for large molecules, but would be only a rough approximation in the short chain hydrocarbons. From the known values of the latent heats of evaporation, the values of γ can then be calculated. Practically a ll of the normal hydrocarbons, with the exception of methane, give the value $\gamma = 34 \pm 1$ ergs $cm.^{-2}$. This value is of the same order of magnitude as the value 48 found from measurements of surface tension. The fact that γ is found to be constant proves that the work done in removing molecules from the liquid to the vapour phase is strictly proportional to the molecular surface. In other words, the latent heat of evaporation is proportional to the two-thirds power of the molecular volume.

This theory can readily be extended to cover the case of the heats of evaporation of the various aliphatic alcohols. If S is the total surface of the molecule of vapour, then aS is the surface of the head of the molecule (hydroxyl group) while cS is the area of the tail (hydrocarbon chain). We may let γ_a and γ_c represent the surface energies per unit area of the heads and tails respectively when the molecule is in the vapour phase. Thus $Sa\gamma_a$ is the total energy of the head and $Sc\gamma_c$ the total energy of the tail.

If the molecules of alcohol in the liquid phase were arranged at random, that is, if they did not orientate each other appreciably, and did not tend to form clusters (segregation), then it may readily be shown that the total interfacial surface energy in the liquid between the hydroxyl groups and the hydrocarbon chains will be $Sac\gamma_{ac}$, where γ_{ac} is the interfacial surface energy per unit area. We should therefore be able to calculate the latent heat of evaporation, λ, from the difference between the energies in the vapour and liquid phases.

$$\lambda = S\,(a\gamma_a - ac\gamma_{ac} + c\gamma_c)$$

If we take $\gamma_a = 193$, $\gamma_c = 34$, and $\gamma_{ac} = 34$, we get excellent agreement between the observed latent heats of evaporation and the structures of most monobasic alcohols. We have already seen that γ_c is equal to 34 as the value found for the evaporation of pure hydrocarbons. The value $\gamma_{ac} = 34$ was found as a result of experiments on the vapour pressures of mixtures of alcohol and water. The interfacial energy between water and a hydro-carbon is about 59, so that the value 34 is of the right order of magnitude. The surface energy of water is 117, but this naturally represents the surface energy of the least active part of the water molecule, whereas the energy γa equal to 193 corresponds to the most active part of the hydroxyl group so that this value also appears reasonable.

In the case of the alcohols having very long hydrocarbon chains another effect can be clearly seen from the experimental results. In the molecule of vapour the hydroxyl group is able to bury itself at least partially among the coils of the hydrocarbon tail, so that the surface energy of the vapour molecule decreases considerably, beginning with chains of five or six carbon atoms in length. With shorter chains than this, the hydroxyl group is probably fully exposed. Particularly interesting results which are in good agreement with the principle of independent surface action are found in the case of dibasic and tribasic alcohols. The heat of evaporation depends to a great extent on whether the separate hydroxyl groups are able to come into contact with each other in the molecule of vapour and thereby decrease the surface energy.

The theory can readily be extended to calculate the partial vapour pressures of binary solutions. The complete theory, taking into account orientation and segregation of molecules within the liquid, would be very complicated, but in many cases where the forces between molecules are not too strong, these effects can be neglected in a first approximation. Assuming then a random orientation and distribution of the molecules, the total surface energy per molecule in a solution of any given concentration in terms of interfacial surface energies, such as φ_{ac}, and surface fractions, such as a and c, the work done in transferring a molecule from a liquid to a vapour phase can then be calculated, and thus by applying the Boltzmann equation, it is possible to calculate the deviations from Raoult's law. Thus, the partial pressure of any liquid in a binary mixture is given by

$$p\mathrm{A} = AP_{\mathrm{A}} \exp (\varphi S_{\mathrm{A}} ß^2/kT)$$

where p_{A} is the partial pressure of the substance A, the total pressure of the pure component A is P_{A}; the mole fraction of the component A is represented in this equation by A. The quantity S_{A} is the surface area per molecule of area; φ is a constant characteristic of the binary mixture but independent of the concentration of the components, which can be calculated in terms of such quantities as a, c, γ_{ac}, etc. The quantity ß may be called the surface fraction of the component B in the binary mixture of A and B. It corresponds to the ordinary conception of mole fractions but is expressed in terms of the relative surfaces of the molecules instead of the numbers of the molecules. Thus

$$ß = BS_{\mathrm{B}}/(AS_{\mathrm{A}} + BS_{\mathrm{B}})$$

This equation with only one adjustable constant, φ, apparently agrees in general better with experimental data than a somewhat similar equation with two adjustable constants derived by van Laar on the basis of thermodynamical considerations. C. P. Smyth has recently used this equation in connection with his own measures of the vapour pressures of binary mixtures. The agreement, in most cases, is fairly good, but, as is to be expected in the case of more polar molecules, mixtures of alcohols with water show considerable deviations. It is probable that these can in large part be taken into account by developing the theory further to allow for orientation and segregation of the molecules within the liquid. Views of the type which I have been discussing may thus be applied quantitatively, often with considerable accuracy, in studying the interactions between molecules of organic substances. The energy relations based on the conception of surface forces between molecules, together with the Boltzmann equation, frequently permit decisions to be made as to the mechanism of various surface phenomena.

For example, in expanded films of oils on water it has often been assumed that the molecules could remain erect on the surface without touching one another. Simple energy considerations of the kind which we have been using indicate immediately that this is impossible. The tails of the molecules must remain in contact with each other in the case of long chains, but with shorter chains, the molecules may separate but must then lie flat upon the surface of the water. The principle of independent surface action affords one of the greatest safeguards against the setting up of impossible hypotheses.

Chapter 3

Analytical Chemistry

There is general agreement among historians that the early beginnings of chemistry were largely founded on a metallurgical basis, and that a certain amount of knowledge of a few metals, particularly of the precious metals gold and silver, existed in very early times. The story of the discovery by Archimedes (212 B.C.) of the fraud perpetrated by a goldsmith in partially substituting silver for gold in a crown ordered by King Hiero of Syracuse is well authenticated, and the principle of the method which he devised-the determination of specific gravity—has become of immense practical value in many directions. The processes of assaying precious metals by cupellation, and of testing their degree of fineness by examination of the streak left on a piece of rock, later known as a 'touchstone', when a fragment of the metal is drawn over the surface, are of very ancient date. Geber, described by more than one writer as the greatest of the Arabian alchemists, who lived in the eighth century A.D., knew that the precious metals resist oxidation when heated in air, whereas lead increases in weight when subjected to this treatment. Indeed, Geber has left descriptions of the method of obtaining gold and silver by cupellation which bear comparison with the dry methods of assaying these metals at the present time.

The action of nitric acid in 'parting' alloys of gold and silver was also known to Geber, but, according to Roberts-Austen, the first official mention of its use in assaying was in 1343 in the work of the French Mint. The use of the balance in assaying has been practised for centuries; and again to quote Roberts-Austen' the use of the balance in very early times, for the purpose of assay, absolutely demolishes the claim of quantitative chemistry to be considered of comparatively modern origin'. As chemistry gradually evolved from alchemy, and as experimenters became familiar with an increasing number of substances, means for their recognition and classification were gradually evolved. In the eighteenth century, a period of great literary and scientific activity, some methods of considerable value in chemical analysis, both qualitative and quantitative, were devised.

At the beginning of that century, Newton assumed the office of Master of the Mint, having occupied the office of Warden some three years previously. At the outset he rapidly mastered the technique of assaying with the object of directing the work of his assistants, but he appears to have had no great regard for metallurgy as a science; indeed, he remarked that 'the assay master acts only as a manual artificer', and again, 'refining and assaying are manual trades'. Of greater interest are the researches of those who devoted their attention to experimental work in pure chemistry, and thereby devised analytical methods in the course of their investigations.

Thus Black in his work on *magnesia alba*, which was published in 1755, worked out a gravimetric method for estimating the *fixed air* liberated from a carbonate by the action of heat or of an acid. Some eleven years later Cavendish determined the carbon dioxide displaced in reactions of this kind by the method of loss of weight.

Figure: *The presence of copper in this qualitative analysis is indicated by the bluish-green color of the flame.*

Qualitative analysis of minerals with the aid of the blowpipe appears to have originated with Cronstedt, and shortly afterwards was greatly extended by Bergman, who published a work *entitled De tubo ferruminatorio* in 1779. The value of the blowpipe as a means of obtaining information regarding the qualitative composition of minerals was thoroughly recognised by Berzelius, who published a book on the subject, an English translation of which appeared in 1822, entitled *The Use of the Blowpipe in Chemical Analysis and in the Examination of Minerals.* This work contains minute directions for carrying out experiments of this kind, together with clear descriptions of the behaviour of some thirty substances according as they are heated in the oxidising flame or in the reducing flame in the presence of various fluxes, such as sodium carbonate, borax and microcosmic salt. The blowpipes in use at that time were usually operated in conjunction with candle flames or with the flame of an oil lamp; but spirit lamps were commonly used for observing the effect of heat upon substances in glass tubes; indeed, in numerous chemical operations the spirit lamp was very generally employed as a source of heat until the invention of the well-known Bunsen burner about the year 1855.

Bunsen showed that most of the dry reactions of inorganic substances, which were effected with the aid of the blowpipe, could be carried out directly in the flame of his burner by careful attention to placing the substance under examination in the right zone of the flame according as an oxidising or reducing action is required. Although analysis by dry methods was brought to a high degree of refinement by Bunsen, it is seldom employed in so elaborate a manner at the present time. Nevertheless, it is generally recognised by modern analysts that a preliminary examination of a substance by dry methods, particularly by the aid of borax bead tests and flame colourations, is of much value in modifying or even determining subsequent procedure.

The development of inorganic analysis, both qualitative and quantitative, was intimately connected with the study of minerals, and the influence of Berzelius and of some of his immediate followers—Rose, for example—extending from the very early years of the nineteenth century to much later times was far-reaching. The experimental work was difficult, often very difficult, but the skill shown in devising methods for the detection, separation and gravimetric determination of the elements, including some described by certain modern writers as 'rare' elements, had a profound influence on the whole subsequent development of inorganic chemistry. An important practical *Handbuch der analytischen Chemie* by Rose in two volumes, one qualitative and the other quantitative, first appeared in 1829, and by 1834 had reached a third edition. Even at that early date descriptions are to be found of operations which are still in use.

These include the familiar division of the metals into analytical groups, the use of suspensions of barium carbonate for effecting the separation of the very weakly basic hydroxides from the less weakly basic ones, the use of succinic acid for effecting the separation of iron from 'manganese, and the use of tartaric acid for avoiding the precipitation of certain metals in alkaline solution. The *Handbuch* of Rose, after passing through several editions, appeared in an enlarged and revised form with the title of *Ausführliches Handbuch der analytischen Chemie* in 1851, and a number of references to the valuable methods there described are to be found in Mellor *Treatise on Quantitative Inorganic Analysis*, which was first published in 1913. The standard work on *Applied Inorganic Analysis* by Hillebrand and Lundell, which appeared in 1929, contains a remarkable tribute to 'the great workers in chemistry of the earlier half of the nineteenth century, and for the painstaking care and accuracy of Berzelius, Wöhler and others, the mineralogists and geologists of today have reason to be thankful. Considering the limited facilities at their disposal in the way of laboratory equipment and quality of reagents, the general excellence of their work is little short of marvellous.'

As is well known, the analysis of minerals resulted in the discovery of some previously unknown elements, and this soon led to the determination of their atomic weights as well as to the revision of those of elements already known. Some of the values obtained by older workers will bear comparison with those obtained by more modern investigators, and a very interesting example is to be found in the case of boron. In 1822 Berzelius determined the water of crystallisation in borax from which he obtained a value for the atomic weight of 11·08. Two years later he converted anhydrous borax into sodium sulphate, from which a figure of 11·0 was obtained.

In 1893 Ramsay and Aston, using the original method of Berzelius, gave a value of 10·94. They also converted anhydrous borax into sodium chloride by repeated evaporation with hydrochloric acid and methyl alcohol, which was really an adaptation of a method used by Penny in 1839 by which he obtained results of great accuracy for several equivalent numbers, and thus obtained a figure of 10·96. Modern atomic weight determinations are nearly always made by the silver-halide methods elaborated chiefly by Stas and by Richards, and a redetermination of the atomic weight of boron by Briscoe and Robinson in 1925, in which the ratio of boron trichloride to silver was used on minerals containing the element obtained from widely different parts of the world, gave values of the atomic weight between 10·82 and 10·84.

A highly interesting example of the value of research in inorganic analysis is to be found in the chemistry of niobium and tantalum, the oxides of which are sometimes known as the earth-acids in order to contrast them with those of the rare earths which are strongly basic. Many compounds of niobium and tantalum are of ill-defined composition, and most of them show a marked tendency to undergo hydrolysis and to give rise to colloidal precipitates. In 1801 Hatchett analysed the mineral columbite and obtained evidence of the existence of a new element which was then termed columbium, and in the following year Eckeberg claimed to have discovered an element which he called tantalum. But, as Schoeller has pointed out, neither Hatchett nor Eckeberg was aware of the precise nature of the earth-acids, which were undoubtedly mixtures; it is therefore an error to ascribe the discovery of columbium, now usually termed niobium, to the former investigator, or of that of tantalum to the latter.

Between 1809 and 1853 these earthacids were the subject of investigation by Wollaston, Berzelius, Wöhler and Rose, and the results of their several labours appeared to indicate the existence of two distinct elements, though not one of them succeeded in separating the two earth-acids in a state of even approximate purity. No real progress was made until the year 1865 when Marignac was able to characterise niobium and tantalum as separate, though closely similar, elements. He also determined the composition of their alkali salts and complex fluorides and assigned correct formulae to them based on the quinquevalency of the two elements, which was established at about the same time by Sainte-Claire Deville and Troost from determinations of the vapour densities of the anhydrous chlorides. Marignac devised a method of separating the two earth-acids, which, though not perfect, was a marked advance on anything which had been accomplished hitherto. The process was based on the much greater solubility of potassium oxyfluoroniobate, K_2NbOF_5, as compared with that of potassium fluorotantalate, K_2TaF_7. Tantalum is more basic than niobium, and the resulting complex salt is not perceptibly hydrolysed under the working conditions. Determinations of the atomic weights of the two elements by Marignac, in which various ratios were used, yielded values of 94·0 for niobium and 182·0 for tantalum. Having regard to the dates at which these atomic weights were determined—1865 and 1866—the results are in reasonably good agreement with the modern values—92·91 and 180·88 respectively.

The analytical chemistry of niobium and tantalum was greatly advanced as a result of the researches of Schoeller and his collaborators since 1921. In 1925 they showed that tannin was a valuable precipitant for the earth-acids from their solutions in certain weak acids, or of their salts, particularly in tartrate and in oxalate solutions, the last named being used for separating niobium and tantalum from each other. Admittedly the chemistry of these operations is highly obscure. On this subject little can be said beyond recognising—what had been known long before—that tartrates readily form complex ions with numerous metallic hydroxides which thereby prevent their precipitation by alkalis; but in what manner tannin can effect precipitation from complex niobo-oxalates and niobo-tartrates is much more difficult to follow. The precipitates have been described, no doubt correctly, as adsorption complexes, as opposed to compounds of definite stoichiometrical composition. The separation of the two earth-acids from each other by a method based on the differential stability of oxalotantalic and oxaloniobic acids in the presence of tannin in slightly acid solution was described in detail by Schoeller in 1930. Fractional precipitation was necessary, but could be followed by the difference in colour of the two adsorption products, that of tantalum being yellow, while the niobium complex is red. There can be no doubt that this series of investigations constitutes a major advance in modern analytical chemistry.

The progress in analysis since the time of Berzelius is to be attributed to more than one cause, but particularly to the discovery of new reactions and to the introduction of new reagents, especially organic compounds, many of which are highly sensitive. Microchemistry as such is no new subject. Not infrequently has it happened that some chemical change, such as the production of some visible effect such as a striking colour, has resulted first in its application as a qualitative test and sometimes afterwards in its elaboration into a colourimetric or gravimetric method for quantitative work. A few examples may be quoted by way of illustration.

The celebrated Marsh test for arsenic was devised by its discoverer in 1836, and has been the subject of numerous investigations since that time with the object of ascertaining its sensitiveness and of developing it quantitatively. In 1901 Wanklyn published a monograph on arsenic in which the value as well as the few limitations of this test was discussed in detail. Wanklyn had no hesitation in describing the Marsh test as superior to another test devised by Reinsch in 1842. The test is of great importance in forensic work; and for quantitative purposes, methods have been devised depending on comparison of the arsenical mirrors obtained when the arsine is decomposed by heat with those obtained under parallel conditions from a standard solution containing arsenic, and also depending on the reducing action of the gas upon silver nitrate.

It has long been known that curious complex compounds can be obtained by the action of ammonia on mercuric salts, and the well-known yellow product which is formed by the action of traces of ammonia on an alkaline solution of potassium mercuric iodide appears to have been discovered by Nessler in 1856.

The compound which is formed is known as the iodide of Millon's base, and is now formulated as $[OHg_2NH_2]I$. For the purpose of determining traces of ammonia salts in water colourimetrically, the Nessler reagent has been found of the greatest value, but special directions are necessary for preparing the solution in order to obtain reliable quantitative results.

The production of an intense yellow colour when hydrogen peroxide is added to an acidified solution of a titanium salt appears to have been first observed by Schönn in 1869, and in 1882 Weller elaborated a colourimetric method for the determination of titanium based on this reaction. The value of the method for determining small quantities of titanium has since been fully recognised. The production of the colour was formerly attributed to oxidation of the titanium to the sexivalent condition, corresponding to the oxide TiO_3, since reducing agents destroy the colour, but of late it has been regarded as due to the formation of a complex acid to which the formula $H_2[TiO_2(SO_4)_2]$ has been assigned.

In this, as in other colourimetric methods, attention has to be given to conditions, and especially to securing the absence of substances other than the one to be determined, which might give rise to similar colours. Now many substances which can exist in more than one stage of oxidation are liable to exhibit colours; thus vanadium gives rise to an orange-coloured peroxide derivative under similar conditions to the colour produced by titanium.

An ingenious method for determining the two elements colourimetrically was devised by Mellor in 1912. Fenton in 1908 had shown that dihydroxymaleic acid produced a strong orange-yellow colour with titanium compounds, which he estimated to be 'some fifteen to twenty times as delicate as the hydrogen peroxide test' and 'it is not given by vanadium'. Mellor's method for estimating the two elements together consisted in dividing the very dilute acidified solution into two parts, treating one with hydrogen peroxide and the other with Fenton's acid, and making the necessary observations with the aid of a colourimetre.

The production of an intense scarlet colour or precipitate by adding an alcoholic solution of dimethylglyoxime to an ammoniacal solution of a nickel salt was observed by Tschugaeff in 1905. In the following year it was applied in qualitative analysis by Kraut, and one year later by Brunck in gravimetric analysis. A considerable literature on this subject has arisen since that time, and the method has been recognised generally as of great value—it is, indeed, regularly used for the determination of the element in nickel steels. The scarlet compound is regarded as a co-ordination compound of nickel to which the formula has been assigned.

In a number of directions a physico-chemical approach to various operations, which have long been practised in chemical analysis, has provided the means of obtaining a better understanding of their mode of action, and sometimes of effecting improvements on them. This has been particularly evident with regard to operations in which the degree of acidity is concerned. Reference to the directions for effecting certain separations, as set forth by the older experimenters, has disclosed what may fairly be described as real craftsmanship; and when these descriptions are considered in the light of modern views, it is frequently surprising how closely they correspond to what might have been expected from an approach directed from a more developed theoretical background. Thus the older methods for effecting the precipitation of certain metallic hydroxides while leaving others in solution, by various devices for limiting the precipitating powers of the basic substances which are added, are now discussed in terms of precipitating within the appropriate limits of hydrogen-ion concentration.

The use of suspensions of carbonates or oxides in work of this kind, recommended by the older analysts, has received favourable notice by modern writers, of whom mention should be made of Hillebrand and Lundell.

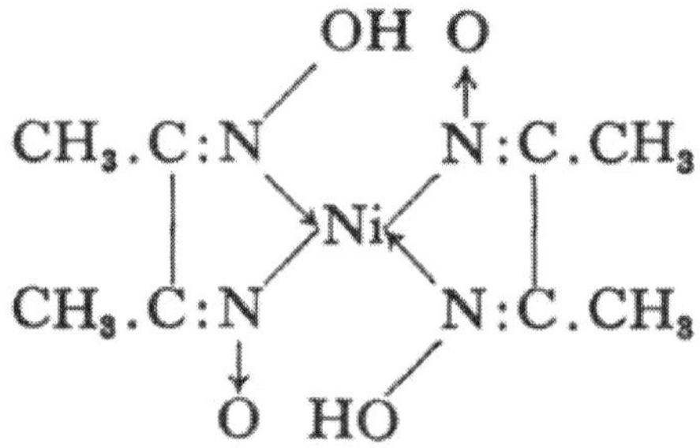

These authors have given careful attention to the pH values of a number of suspensions of carbonates and oxides and have correlated these values with the pH ranges between which certain hydroxides are and others are not precipitated, so as to indicate which are most suitable for effecting any particular separation. This subject has also been studied by a potentiometric titration method by Britton in 1925. Much assistance in this kind of work has also resulted from knowledge of the pH range of indicators, a subject to which much attention has been given in modern times.

Volumetric Analysis

The earliest recorded example of the measurement of alkalinity is to be found in a work on bleaching by Home, published in 1756, in which the 'value' of pearl ashes (potassium carbonate) was estimated by counting the number of teaspoonfuls of dilute nitric acid which were required to be added until effervescence ceased. In 1806 Descroizilles estimated acids and alkalis by noting the volume of one liquid required to neutralise the other, neutrality being determined by using an infusion of violets to serve as an indicator.

A few other examples of experiments of a similar kind are to be found in the works of others. The foundations of volumetric analysis, however, as generally understood, were undoubtedly laid between 1824 and 1832 by Gay-Lussac. His first experiments were concerned with the determination of the value of bleaching powder by employing its decolourising action on a solution of indigo. He also worked on the quantitative neutralisation of acid by alkalis. But his most important contribution to this branch of analysis was published in 1832 under the title *Instructions sur l'essai des matiεres d'argent par la voie humide. Gay-Lussac's method of adding a solution of sodium chloride to a solution of silver dissolved in nitric acid, under strictly specified conditions of working, and ascertaining when precipitation no longer takes place, constitutes the 'wet' assay of silver; it has been regularly practised, with only minor modifications, in most mints ever since that time. Some twenty years after Gay-Lussac devised the method, Mulder investigated the conditions in the vicinity of the equivalence point, when the solution is saturated with silver chloride, the solubility of which though very small is nevertheless perceptible, and laid down the conditions for working the method to the best advantage.*

Recognition of the value of volumetric methods was shown by the appearance of practical books on the subject. MohrLehrbuch der chemisch-analytischen Titriermethode was published in 1855, and in 1863 was followed by the publication of Sutton Systematic Handbook of Volumetric Analysis. Both works appeared at various later dates in subsequent editions. On the experimental side Mohr made a useful contribution by introducing potassium chromate as an indicator for titrating neutral chlorides by silver nitrate.

The method is much more rapid than the original procedure of GayLussac, but of an inferior degree of accuracy. In 1874 an important method for determining silver in acid solution by titration with gammonium or potassium thiocyanate was introduced by Volhard, in which the end-point is indicated by the use of ferric sulphate. The method has been very widely used since that time, not only for the determination of silver in alloys, but also for the estimation of halides by residual titration of the excess of silver by standardised thiocyanate. The determination of chlorides by this method was at first found to be inaccurate, until it was recognised that removal of the precipitated silver chloride was necessary before proceeding to the residual titration. The influence of studies on the solubilities of very sparingly soluble substances is very relevant in this connexion. By physico-chemical methods it has been shown that silver chloride is more soluble than the thiocyanate, the bromide is of nearly equal solubility, and the iodide is the least soluble of the three halides.

As ordinarily practised in mints the Gay-Lussac method for testing the purity of fine silver is capable of realising a limit of accuracy of about 0·2-0·3 part per thousand. This is a considerably higher degree of accuracy than is possible by the Volhard method as ordinarily carried out. In 1900, however, T. K. Rose investigated the possibilities of the Volhard method, and found that when worked under certain specified conditions it could be made to yield results which surpassed the Gay-Lussac procedure in refinement, and he was able to realise a limit of 0·1 part per thousand; but the process was too slow to be of much practical value in routine assay work.

A new departure in argentometric work arose in 1923 by the introduction of adsorption indicators by Fajans. By using certain dyestuffs, such as fluorescein and its derivatives, the titration of halides by silver nitrate can be carried out with a degree of accuracy not inferior to what can be realised by the Mohr process or by the Volhard method. The use of the fluorescein class of indicators is limited to work in neutral solution. Extension of this work by Fajans and by others has led to the use of other dyestuffs which can enable such titrations to be carried out in acid solution, particularly tartrazine and phenosafranine. By taking advantage of the difference in solubility and of capacity for adsorption of silver chloride and silver iodide, it has been found possible to determine a soluble chloride and an iodide in the same solution by differential titration.

Gay-Lussac made some use of the reducing action of a solution of arsenious oxide to determine the value, or, as it is usually termed, the available chlorine, of bleaching powder. He expressed his results in what were termed Gay-Lussac degrees. These units indicate the number of litres of chlorine, reduced to 0°C. and 760 mm., obtainable from 1 kg. of the material; 100 of such degrees are equivalent to 31·8% of available chlorine. Gay-Lussac's arsenite method was improved by Penot in 1852, and is still in general use.

Several other volumetric methods of analysis were developed about the middle of the nineteenth century. Thus in 1846 Marguerite devised the well-known method of titrating acidified ferrous salts with potassium permanganate. This was followed in 1850 by Penny's potassium dichromate method for estimating iron. Mohr introduced ferrous ammonium sulphate—sometimes known as Mohr's salt—for standardising these oxidising agents. Iodometric methods were devised by Bunsen in 1853, and described in detail in a paper entitled *Ueber eine volumetrische Methode von sehr allgemeiner Anwendbarkeit.*

Bunsen's research on iodometry began with a thorough investigation of the reversibility of the reaction between iodine and sulphurous acid. It had previously been used as a means of determining iodine by titration with a dilute solution of the reducing agent by Dupasquier; but Bunsen was able to show that the oxidation to sulphuric acid, with simultaneous reduction of the iodine to hydriodic acid, is only quantitative if the concentration of the sulphur dioxide does not exceed 0·04% by weight. Bunsen developed methods for determining a number of oxidising agents by quantitative liberation of iodine from acidified potassium iodide and for determining the iodine by titration under strictly specified conditions. Just after the publication of Bunsen's paper, Schwarz introduced sodium thiosulphate as a substitute for sulphurous acid in iodine titrations, and this reagent very soon came into general use on account of its much greater stability.

From the earliest beginnings of volumetric analysis, the determination of acids and alkalis by reciprocal titration has been a prominent feature of this kind of work. Among the various natural products which have been used as indicators in acidimetry and alkalimetry, litmus has long enjoyed an exceptional degree of popularity, and even at the present time its use has by no means been abandoned. But its limitations, particularly its sensitiveness to carbon dioxide, soon became recognised; and the introduction of methyl orange by Lunge in 1881 for titrating alkaline carbonates was an important practical advance. R. T. Thomson in 1883 made an exhaustive study of the various indicators in common use at that time, and pointed out the best practical conditions for applying them to any particular determination. In 1891 Ostwald published his well-known theory of indicators based on the principle of ionic equilibria.

Although this theory is no longer accepted in the simple form presented by its author, and has been modified by the introduction of Hantzsch's views on pseudo-acids and pseudo-bases, Ostwald's ideas have been of very great value in obtaining a clear understanding of the mode of action of these substances and of making the right choice of the most suitable indicator for any particular titration. Since 1904 it has been shown by Salm, Thiel and others that the colour changes of indicators is determined by the actual concentration of hydrogen ions in solution, and this can be given numerical expression. In 1911 Tizard published a report on the sensitiveness of indicators, in which he drew attention to the effect of the amount of indicator on the end-point of a titration, according as a two-colour indicator, such as methyl orange, or a one-colour indicator, such as phenolphthalein, is used. A very important monograph entitled *Die Theorie der alkalimetrischen und azidimetrischen Titrierungen* was published by Bjerrum in 1914. In this work Bjerrum discussed some of the topics, to which reference was made in Tizard's report, in more elaborate detail, and dealt very thoroughly with problems such as the degree of hydrolysis of salts.

He also adverted to the possibility of effecting further advances in this department of practical chemistry, in particular to titrating to any specified hydrogen-ion concentration. This has been realised experimentally in two ways, namely, by potentiometric titration and by using indicators having a colour change corresponding to the desired pH interval. The number of indicators, the pH range of which is accurately known, is now fairly large; and of these the sulphonephthalein series is of particular importance. This class of indicator has received much attention at the hands of Acree and his collaborators since 1916; and it may be noted, in passing, that certain sulphonephthaleins, such as thymol blue, have a double colour change, each corresponding to a definite pH interval. The existence of such a double colour change was explained by White and Acree in 1919 by regarding the sulphonephthaleins as dibasic acids, the primary salt being derived from the carboxylic acid resulting from the opening of the lactone ring and the secondary salt from one of the phenol groups.

Much progress in problems connected with the estimation of acids and alkalis has resulted from a correct understanding of the significance of the ionisation constants of these classes of compounds including the physico-chemical properties of indicators. Two examples may be quoted by way of illustration. The accuracy attainable in the titration of mixtures of acids and bases, and the titration of polybasic acids, was discussed by Tizard and Boeree in 1921. They showed that dibasic acids can be titrated as monobasic if there is a wide difference between the values of the two ionisation constants, but not when this difference is small; thus by using an indicator of the methyl orange type, maleic acid can, and succinic acid cannot, be titrated to the first stage. The estimation of ampholytes, particularly amino-acids, has been the subject of numerous investigations. In this kind of work the problem has been one of eliminating the effect of the basic constituent, and two methods for effecting this object have been devised and have come into general use. Sörensen in 1907 devised a displacement method which consisted in using formaldehyde to react with the amino-acid to form an extremely weakly basic derivative, which thus enables the acid to be titrated with alkali and phenolphthalein. The other method depends on the effect of alcohol in exerting a lowering effect on the strengths of amino-acids, but this effect is much more pronounced on the basic constituent. This method was introduced independently by Foreman in 1920 and by Willstätter and Waldschmidt-Leitz in 1921, and consists in titrating the amino-acids directly with alkali in a solution containing a very high concentration of alcohol and using a sensitive indicator such as phenolphthalein.

During the twentieth century several new volumetric methods depending on reactions involving oxidation and reduction came into existence. In 1903 Knecht introduced titanous chloride as a quantitative reducing agent for a number of determinations, particularly for estimations of certain classes of dyestuffs. On account of its very powerful reducing properties, titrations with titanous chloride require to be carried out in an atmosphere of hydrogen. Later it was found that the sulphate is more stable, and accordingly does not require such close protection from atmospheric oxidation. In 1903 Andrews showed that the reaction between an iodide and an iodate in acid solution can be made to take a wholly different course from the usual one in which one equivalent of the iodate reacts with five equivalents of iodide. He found that when a solution of potassium iodate is added to one of an iodide in presence of a very high concentration of hydrochloric acid, the iodine which is first liberated is oxidised to iodine monochloride, one equivalent of the iodate reacting quantitatively with two equivalents of iodide, and he elaborated a very valuable method depending on this principle. Andrews's method of direct titration with potassium iodate has been applied in a number of directions, for example, by Berry in 1926 for the determination of thallous salts. Another modern volumetric oxidising agent, namely, ceric sulphate, was introduced in 1928 by Furman and independently by Willard and Young. Although the powerful oxidising properties of this substance had long been known, and a few instances of its use in volumetric work had even been published, these authors demonstrated its great value and wide applicability. Broadly speaking, it can be used for all determinations which are commonly effected with acidified potassium permanganate, as well as for some for which that reagent is not altogether suitable.

The completion of the reaction of any volumetric method should, whenever possible, be indicated by some striking colour change within the liquid. The older oxidation methods, with the important exception of Penny's dichromate method which requires an external indicator, satisfy this requirement. But it has long been known that certain aromatic compounds can undergo reversible oxidation and reduction from the benzenoid to the quinonoid condition, the process taking place with well-marked colour changes, and in 1924 Knop utilised diphenylamine as an internal indicator for titrating ferrous salts with dichromate. The introduction of some of the newer volumetric oxidising agents, especially ceric sulphate, has led to renewed studies directed with the object of discovering the most suitable substances to serve as internal indicators. In this connexion the oxidation potential of the indicator is of great importance. In 1931 Kolthoff and Sarver found the diphenylamine sulphonic acid was preferable to diphenylamine, and this derivative in the form of the barium salt has come into general use. Of the other substances which have been used as internal oxidation reduction indicators, mention should be made of triorthophenanthroline-ferrous sulphate which was first employed in 1931 by Walden, Hammett and Chapman. They claimed that this substance fulfilled the requirements of an ideal indicator on account of its high oxidation potential, excellent reversibility of the colour change, and stability in solution.

Gas Analysis

The subject of gas analysis should be dated from the latter years of the eighteenth century, and the name of Cavendish will always be associated with the first accurate measurements of the proportions of the two chief constituents of the atmosphere. In a paper entitled *An account of a new eudiometre* which appeared in 1783, Cavendish elaborated a method, founded on a discovery made by Priestley some eleven years previously, for estimating the proportion of dephlogisticated air (oxygen) in the atmosphere. This consisted in causing nitrous air (nitric oxide) to react with the oxygen, dissolving the soluble product of the reaction in water, and measuring the resulting contraction in volume. These experiments were attended with considerable difficulty in attaining reproducible results, because it was not at that time understood that nitric oxide can combine with oxygen in varying proportions. Cavendish, however, obtained consistent results by adding the air to the nitric oxide and measuring the volume which disappeared during contraction.

According to Cavendish, 'the whole diminution in mixing is equal to the bulk of nitrous air, which is turned into acid, added to the diminution which the common air suffers by being phlogisticated'. In this statement the expression *being phlogisticated* meant what would in modern phraseology be termed *being deprived of oxygen*. Cavendish thus appreciated, what was later more clearly understood by Dalton and by Gay-Lussac, that oxygen can combine with twice or with four times its volume of nitric oxide; but, provided that the latter gas is in excess, the volume of oxygen withdrawn is constant. The accuracy which Cavendish attained by this method, long since abandoned, bears favourable comparison with that of modern experimentalists using more refined methods. The celebrated paper, entitled *Experiments on air*, which was published by Cavendish in 1784, contained an account of his determination of the combining ratios of hydrogen and oxygen by explosion, and may be regarded as marking the foundation of that method of analysis.

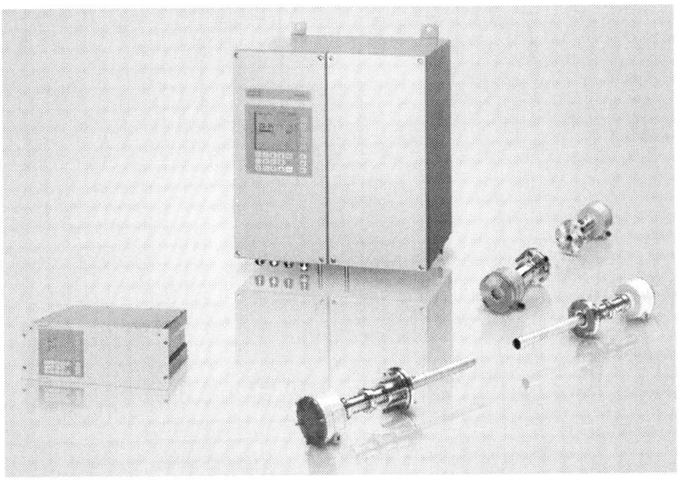

Figure: *Gas Analyzer*

In 1790 Volta constructed a eudiometre for determining the composition of combustible gases by explosion with oxygen, and this apparatus was also used by others, notably by Dalton in 1804, who ascertained the volume composition of marsh gas (methane) and of olefiant gas (ethylene) by sparking and measuring the resulting contraction in volume. It will be appreciated that correct formulae could not at that time be assigned to these gases as Avogadro's hypothesis had not yet been formulated; but the results are of historic interest as having occupied Dalton's mind in connexion with his law of multiple proportions. Analysis by sparking has been widely used since that time, notably by Bunsen and by later experimentalists. Coming to more recent years, the method was used by Diels and Wolf in 1906 for determining the molecular composition of carbon suboxide, a compound discovered by them to which they assigned the formula C^3O^2.

Measurements in the analysis of gases are, broadly speaking, apart from noting contractions resulting from the condensation of steam after explosion, effected by the application of selected reagents, either liquid or solid, and observing the resulting diminution of volume, or, alternatively, by weighing. Analysis by absorption using solid reagents was brought to a high degree of refinement by Bunsen, and described in a work entitled *Gasometrische Methoden*, an English translation of which, by Roscoe, appeared in 1857. Greater rapidity of working, although of a somewhat inferior degree of accuracy, was secured by the introduction of liquid reagents a few years later and the invention of more convenient apparatus, especially the forms devised by Hempel in the latter years of the nineteenth century.

For the present purpose it is unnecessary to refer in detail to the various reagents which are commonly employed in the analysis of gases by absorption, but a few remarks may be added on the use of palladium for the estimation of hydrogen. The property possessed by this metal of occluding large volumes of hydrogen was discovered by Graham in 1866, and the nature of the material thus obtained-whether a true hydride or an adsorption product—has been the subject of numerous later investigations. It appears to have been first applied analytically by Keiser in 1887 in a determination of the atomic weights of hydrogen and oxygen, the hydrogen having been absorbed in palladium and weighed directly. This method was also employed in a more elaborate research on the same subject by Morley in 1895. Bone and Jerdan employed palladium in 1901 for absorbing hydrogen in ordinary gas analysis, but their method of working involved the use of a mercury pump. An important practical advance was made in 1910 by Paal and Hartmann, who introduced the use of colloidal palladium in conjunction with a Hempel gas pipette. Since Graham's time it has been recognised that the rate of absorption of hydrogen is very greatly affected by the physical condition of the metal, the greater the degree of subdivision the more efficient is the occlusive capacity. It is therefore not surprising that palladium in colloidal solution should be a very useful absorbent.

Many analytical processes are carried out by determining the volume of a gas generated in a reaction, and a brief reference to two examples of this kind of work may now be made. In 1847 Crum devised a method 'for the analysis of bodies containing nitric acid and its application to explosive cotton'. He showed that when nitrates are shaken with mercury in the presence of concentrated sulphuric acid, all the nitric acid is reduced quantitatively to nitric oxide, the volume of which is measured directly. In 1878 Lunge devised his well-known nitrometre for the analysis of nitrous vitriol by this means, and this apparatus, which at later dates was improved by him in various ways, has been of much practical value in various other directions.

The determination of urea by oxidation with a solution of sodium hypochlorite to carbon dioxide and nitrogen, and measurement of the volume of the latter gas evolved, was apparently first suggested by E. Davy in 1854, and has been the subject of many subsequent investigations.

It was soon found that more rapid decomposition was effected by using sodium hypobromite, and that the volume of nitrogen liberated was always lower than the calculated value, the deficiency being about 8%. In 1878 Fenton showed that when sodium hypochlorite was used in presence of excess of alkali, only about half of the calculated volume of nitrogen was obtained, whereas with hypobromite all the gas, with the exception of the defect of 8%, was evolved. Fenton traced the deficiency to the formation of cyanate which is not oxidised to nitrogen. These conclusions received general confirmation by E. A. Wemer in 1922.

An interesting example of a problem which was finally solved by a specially devised method of analysis was that of the composition of the hydrosulphites, or, as they have also been termed, hyposulphites. In 1852 Schönbein had observed that solutions having remarkably powerful reducing properties could be obtained by the action of zinc on sulphurous acid, and in 1869 Schützenberger studied this subject at greater length, and obtained salts in an impure condition by causing zinc to react with aqueous solutions of the alkali bisulphites. Schützenberger regarded the acid in solution as having the formula

H^2SO^2, but in 1881 Bernthsen suggested the formula $H^2S^20^4$. On account of the unstable character of the compounds and the great difficulty of obtaining them in anything approaching a state of purity, it was for long difficult to decide between the two formulae. In 1902, however, the question was settled in a particularly elegant manner by Moissan.

Having previously shown that the alkali metal hydrides were represented correctly by the formulae KH and NaH, Moissan demonstrated the formulae of the alkali hydrosulphites by passing sulphur dioxide over potassium hydride and measuring the volume of hydrogen generated. The reaction was found to take place in accordance with the equation:

$$2KH + 2SO^2 = K^2S^2O^4 + H^2,$$

and the formula suggested by Bernthsen was thus verified.

It is a matter of general agreement that great advances in chemical science have resulted from improvements in experimental technique, but in no branch of this subject is this more true than in gas analysis. Special mention may be made of the invention of Sprengel's mercury pump in 1865 for producing high vacua—an apparatus which was used by Graham in his investigations on the selective passage of gases through various materials. At a later date Töpler's mercury pump was in constant use in Ramsay's experiments on the inert gases. The introduction of charcoal cooled to very low temperatures by Dewar in 1904 provided a means of separating the rare gases more simply than was possible by the original method of fractional distillation of liquid air employed by Ramsay and Travers.

Organic Analysis

The determination of the two most abundant constituents of organic compounds, carbon and hydrogen, by complete combustion to carbon dioxide and water respectively, is really a special department of gas analysis. The first beginnings of this subject were made by Lavoisier between 1781 and 1784. His experiments were made by using oxygen for effecting the combustions and potash for absorbing the carbon dioxide, the water being determined by difference. In some unpublished notes which were found after his death, however, it is clear that Lavoisier had made some improvements in his method of working, particularly by using compounds such as mercuric oxide or manganese dioxide for carrying out the combustions and by weighing the water directly after absorption in a hygroscopic salt. It must be added that Lavoisier attained a very limited degree of success with these experiments. The subject was afterwards taken up by other chemists, notably by Gay-Lussac and Thenard and by Berzelius, who made various modifications in the method, notably by mixing potassium chlorate, together with an inert substance to moderate the action, with the organic compound and thus effecting the combustion. Some improvement was, no doubt, realised in this way, but it was not until the year 1831 that a really satisfactory procedure was devised by Liebig, which consisted in burning the substances in a stream of air or oxygen, and passing the products of combustion over heated copper oxide to ensure complete combustion before reaching the vessels for weighing them: a tube containing calcium chloride retained the water and bulbs charged with a solution of potash absorbed the carbon dioxide. It was soon found that irregular results were obtained when the combustible substance contained certain elements, notably nitrogen, which gave rise to oxides of nitrogen during combustion and consequently a fictitious increase in the weight of carbon dioxide; but these difficulties were overcome by using heated copper to decompose the nitrogen oxides. The process as devised by Liebig, except for the mode of heating the tube, has continued practically unchanged to the present time. For upwards of a quarter of a century until after the introduction of gas-fired furnaces, combustions were carried out by Liebig's original type of apparatus, a furnace heated by burning charcoal, and the operation required considerable manipulative skill.

The beginnings of a method for determining nitrogen are to be found in the works of Gay-Lussac and Thenard, but no real progress in this direction was made until Liebig had elaborated his method of determining carbon and hydrogen by combustion. In 1831 Dumas devised his well-known method for determining this element; and it may be said that this arose as a consequence of the irregularities which were observed when nitrogenous compounds were submitted to combustion.

Dumas's method consisted in effecting combustion with heated copper oxide in an atmosphere of carbon dioxide, decomposing any oxides of nitrogen with hot copper, and finally collecting the resulting nitrogen by displacement with carbon dioxide over a concentrated solution of potash. Apart from the introduction of a useful form of nitrometre due to Schiff in 1868, the process as devised by Dumas is, with only minor modifications, still in general use.

As many organic compounds when heated with caustic alkalis evolve any nitrogen which they contain in the form of ammonia, methods of estimating this element by determining the ammonia by volumetric analysis received a certain amount of attention in the earlier years of the nineteenth century, but owing to their somewhat limited applicability, they were less popular than Dumas's method. But in 1883 Kjeldahl devised a valuable method, which consisted in decomposing the compound by heating it with concentrated sulphuric acid, and thus converting the nitrogen into ammonium sulphate. The resulting ammonium sulphate was then decomposed by sodium hydroxide and the ammonia determined volumetrically. The function of the sulphuric acid in the Kjeldahl method is partly one of dehydration and partly that of an oxidising agent. Since the method was first devised, it has been the subject of many investigations directed with the object of extending its range of applicability. At the present time it may be said that the Kjeldahl method, as now practised, gives reliable results with widely different types of organic compounds, and for technical work it is in almost universal use; but for purely scientific investigations the Dumas method is still preferred.

From what has been stated it will be clear that the determination of the elements in organic compounds is a matter of destructive oxidation, whether by 'dry' or by 'wet' combustion, so as to render the element in question available in a form suitable for weighing or for volumetric estimation. These principles are to be seen in the method most commonly used for the determination of the halogens and sulphur, which was first devised by Carius in 1860, and improved by him at later dates. In Carius's method the destructive oxidation is effected by heating with fuming nitric acid in sealed tubes. For the determination of halogens the operation is carried out in the presence of silver nitrate, and the silver halide resulting from the reaction is collected and weighed. When sulphur is to be determined, the silver nitrate is omitted and the sulphuric acid is weighed as barium sulphate.

During the first half of the nineteenth century, many attempts were made to devise a method for determining oxygen directly in organic compounds. Such methods were fore doomed to failure, but the ingenuity which was expended in this direction showed clearly how much a direct method for determining this element was desired. Even as late as the year 1865 Ladenburg devoted attention to this subject, and described an ingenious method for elementary analysis in which the compound was decomposed by destructive oxidation with silver iodate and concentrated sulphuric acid. In this method the oxygen was estimated by an adaptation of Bunsen's iodometric procedure, but it is scarcely surprising that an elaborate method of this kind should never have come into general use. It soon came to be recognised that research of this sort was fruitless, and consequently oxygen is always determined by difference, after all the other elements in the compound have been detected and estimated. An oft-quoted example of the importance of a thorough search for elements other than those merely suspected is the substance taurine. This compound was first discovered by Gmelin in 1826 in the bile of oxen, and analysed twelve years later by Dumas and Pelouze, who assigned the formula $C^2H^7NO^5$ to it. But in 1846 Redtenbacher showed that the compound contains sulphur, and after a more thorough analysis he found the correct formula to be $C^2H^7NO^3S$, the atom of sulphur taking the place of the supposed two atoms of oxygen.

In consequence of the rapid development of organic chemistry it became necessary to supplement the methods of elementary analysis with means of determining the actual number of certain kinds of groupings in the molecules of the compounds, in order to deal with problems of establishing their constitution. The general principle of such methods has always consisted in preparing derivatives of such groupings in the form of compounds, which are themselves readily separable in fairly pure condition and analysable by gravimetric or volumetric methods. Thus compounds, such as carbohydrates which are rich in hydroxyl groups, are easily acetylated or benzoylated, and the resulting acetyl or benzoyl compounds analysed by quantitative hydrolysis. The investigation of many natural products which contain methoxy groups in the molecule has been greatly facilitated by a method devised in 1885 by Zeisel for determining such groups. This method consists in decomposing the compound by heating with concentrated hydriodic acid, which results in separating the methoxy group quantitatively as methyl iodide, and this latter compound is distilled into a solution of alcoholic silver nitrate. The resulting silver iodide is then weighed directly. The original procedure of Zeisel was somewhat elaborate, and has subsequently been improved and simplified, notably by W. H. Perkin (the younger) in 1903.

In every branch of analytical chemistry, much progress has resulted in consequence of the development of methods for dealing with successively smaller quantities of substances. Quantitative microchemistry has become of ever increasing importance, particularly during the last half-century. Much of this progress has arisen in consequence of improvements in the design and construction of balances capable of weighing quantities of material of the order of a few milligrammes with an accuracy corresponding to what was regularly obtainable in dealing with decigramme quantities. Apart from some highly sensitive balances which have been constructed at various times, usually in connexion with special researches, such as the quartz torsion microbalance devised by Nernst in 1903, the modern microbalances used in analysis are really an outcome of improvements of a type used in assaying precious metals. Roberts-Austen in his work on *The Metallurgy of Gold* stated that 'the balance must readily indicate differences of 0·5 per 1000 or 1/40th milligramme with a weight of 1/2 gramme in each pan'.

The practice of modern quantitative micro-analysis has resulted very largely in consequence of co-operation between the chemists and the scientific instrument-makers. Emich had shown that analytical determinations could be carried out with the aid of a Kuhlmann assay balance sensitive to 1/100 mg.; and in 1910, as a result of the influence of Pregl, this type of balance was improved in two directions, namely, in respect of the maximum load and by a tenfold increase in sensitivity. As an example of the value of work on this scale the following figures obtained by Pregl may be quoted. A determination of the percentage of methoxyl in vanillin by Zeisel's method using 3·750 mg. of the substance gave a yield of 5·780 mg. of silver iodide, corresponding to 20·4% of methoxyl in the compound, in perfect agreement with the calculated value. It is no exaggeration to state that much of the modern development of organic chemistry, particularly on the biochemical side, is directly due to the introduction of these microchemical methods.

Adsorption phenomena are of great importance in several directions in analytical chemistry. In some kinds of preparative work the withdrawal of impurities by adsorption with charcoal has long been practised, and an interesting example of an adaptation of this principle arose in unexpected circumstances in 1853. It had been idly, or perhaps wickedly, rumoured in Paris that the bitter taste of ales brewed at Burton-on-Trent was produced by the use of strychnine, and considerable panic arose in consequence of this. Graham and Hofmann were invited to investigate this question. They found that when a known quantity of strychnine was *artificially* added to pale ale, the alkaloid could be completely removed by contact with animal charcoal; it was recovered by filtering and extracting the charcoal with alcohol. Having tested the method, Graham and Hofmann applied it to numerous samples of Burton and other ales, and not a single one was found to contain a trace of the poisonous alkaloid.

Another aspect of this subject must be mentioned briefly, namely, that of chromatographic analysis. The beginnings of this method, which depends upon fractional adsorption, are to be found in some experiments on the separation of the chlorophyll pigments by Tswett in 1906. It was shown that when a solution containing the plant pigments was allowed to flow through a vertical column of a suitable adsorbent, the various constituents are preferentially adsorbed at different levels. By repeatedly washing the adsorbent with the pure solvent, the separated zones become clearly visible. The method has been much elaborated in recent years, and great importance has been attached to making a suitable choice of the solvent and of the adsorbent; it should be added that specially prepared alumina has been found very satisfactory. A great deal of experimental work in analysis has been directed with the object of effecting modifications and improvements in well-known processes. To quote a single example, the literature on the analytical chemistry of phosphorus probably exceeds that of any other element. Although work of this kind is admittedly less spectacular and interesting than investigations with more pretentious objectives, its importance should not be underrated because of the possibility of its ultimate influence on developments in pure and in applied chemistry.

Organotins Disrupt the 11[β]-Hydroxysteroid Dehydrogenase Type 2-Dependent Local Inactivation

Organotins, important environmental pollutants widely used in agricultural and industrial applications, accumulate in the food chain and induce imposex in several marine species as well as neurotoxic and immunotoxic effects in higher animals. Reduced birth weight and thymus involution, observed upon exposure to organotins, can also be caused by excessive glucocorticoid levels. We now demonstrate that organotins efficiently inhibit 11[β]-hydroxysteroid dehydrogenase type 2 (11[β]-HSD2), converting active 11[β]-hydroxyglucocorticoids into inactive 11-ketoglucocorticoids, but not 11[β]-HSD1, which catalyses the reverse reaction. Di- and tributyltin as well as di—and triphenyltin inhibited recombinant and endogenous 11[β]-HSD2 in lysates and intact cells with $[IC_{50}]$ values between 500 nM and 3 [micro]M.

Dithiothreitol protected 11[β]-HSD2 from organotin-dependent inhibition, indicating that organotins act by binding to one or more cysteines. Mutational analysis and 3-D structural modelling revealed several important interactions of cysteines in 11[β]-HSD2. [Cys^{90}], [Cys^{228}], and [Cys^{264}] were essential for enzymatic stability and catalytic activity, suggesting that disruption of such interactions by organotins leads to inhibition of 11[β]-HSD2. Enhanced glucocorticoid concentrations due to disruption of 11[β]-HSD2 function may contribute to the observed organotin-dependent toxicity in some glucocorticoid-sensitive tissues such as thymus and placenta. Key words: cortisol, dibutyltin, 11[β]-hydroxysteroid dehydrogenase, glucocorticoid, inhibition, organotin, toxicity, tributyltin, triphenyltin.

Organotins belong to the most widely used organometallic compounds, with an estimated annual production of approximately 50,000 tons. Derivatives of dialkyltin compounds such as dibutyltin (DBT), diphenyltin (DPT), and dioctyltin (DOT) are used in industry as stabilisers in polyvinyl chloride (PVC) and as catalysts in various products, whereas trialkyltins, including tributyltin (TBT) and triphenyhin (TPT) are used in agriculture as fungicides and pesticides and as antifouling agents for large ships. Organotins are ubiquitous environmental pollutants especially relevant for water ecosystems. Accumulation of these lipophilic compounds has been observed in various species of snails, mussels, and fish, causing an increased incidence of sterility or imposex (imposition of male sex characters onto the female).

The main sources of organotin intake for humans are seafood contaminated because of the exposure to antifouling agents, and drinking water contaminated because of the leaching from PVC water pipes. Additional sources are indoor dust, and liquids stored in plastic containers, including various alcoholic beverages. In higher species, including mammals, organotins tend to accumulate in certain organs, namely liver, kidney, and brain. Organotins efficiently penetrate through the skin and easily cross the placenta and blood-brain barrier.

Comparison of the effects of various tri-alkyltins indicated that the compounds with short alkyl groups such as trimethyltin (TMT) and triethyhin were mainly neurotoxic, whereas organotins with alkyl chains of intermediate length (tripropyltin and TBT) were primarily immunotoxic. The higher trialkyhin homologs trihexyhin and trioctyhin were found to be only slightly toxic; however, further metabolism in vivo converted them to their dialkyltin forms, which are also highly immunotoxic. A single oral dose of DOT, DBT, or TBT induces a dose-related reduction of the relative thymus weight in rats, and impaired cell-mediated immunity was observed after dietary exposure to TPT for several weeks. Furthermore, exposure of pregnant rats to organotins causes reduced birth weight.

Reduced birth weight has also been observed with prolonged intrauterine glucocorticoid exposure. After such an insult, circulating cortisol levels remained elevated throughout adult life, indicating a permanently disturbed regulation of the hypothalamic-pituitary-adrenal axis, which leads to a higher susceptibility for cardiovascular and metabolic disorders including obesity, insulin resistance, and type II diabetes. In the placenta the fetus is protected from the high maternal glucocorticoid concentration through the activity of 11[β]-hydroxysteroid dehydrogenase type 2 (11[β]-HSD2), which converts active 11[β]-hydroxyglucocorticoids (cortisol in human, corticosterone in rodents) into inactive 11-ketoglucocorticoids (cortisone in human, 11-dehydrocorticosterone in rodents). Impaired 11[β]-HSD2 activity, due to mutations or the presence of inhibitors such as glycyrrhetinic acid (GA), strongly correlates with reduced birth weight and metabolic complications in later life of the offspring.

Moreover, exposure of rats to excessive levels of glucocorticoids causes thymus involution, a phenomenon also evident after exposure to organotins. Treatment of rats with high doses of the 11[β]-HSD inhibitor GA led to a significant elevation of systemic glucocorticoid levels accompanied by thymocyte apoptosis. Despite the fact that both exposure to excessive levels of organotins and glucocorticolds cause low birth weight and thymus involution in animal models, the impact of organotins on the control of the intracellular availability of glucocorticoids has not been studied. Therefore, we investigated the effect of various organotins on the activities of 11[β]-HSD1, converting inactive 11-ketoglucocorticoids to active 11 [β]-hydroxyglucocorticoids, and of 11[β]-HSD2, catalysing the opposite reaction. We also studied the mechanism of organotin-dependent inhibition of 11[β]-HSD2.

Chemicals and Reagents

We purchased [1,2,6,7-^{3}H-cortisol, [2,4,6,7-[^{3}]H]-estrone, and [2,4,6,7-[^{3}]H]-estradiol from Amersham Pharmacia; [1,2,6,7[^{3}]H]-cortisone from American Radiolabelled Chemicals; cell culture media and supplements from Invitrogen; and steroid hormones from Steraloids. All other chemicals were obtained from Fluka AG (Buchs,

Switzerland) and were of the highest grade available. Organotins were dissolved in dimethyl sulphoxide (DMSO) and stored as 20 mM stock solutions at -70°C. Human 11[β]-HSD1 and 11[β]-HSD2 expression constructs in pcDNA3 vector (Invitrogen) were described previously. Plasmids containing cDNA from human 17[β]-HSD1 or 17[β]-HSD2, kindly provided by Stefan Andersson, were recloned into pcDNA3 vector by PCR with primers at the 5' end containing a HindIII restriction site (17[β]-HSD1) or a BamHI restriction site (17[β]-HSD2), a Kozak consensus sequence and the initiation codon, and primers at the 3' end containing a stop codon followed by an XbaI restriction site. All constructs were verified by sequencing.

Cell and Organism

It seems to be a necessity of the biological world—though exceptions exist—that any living mass of more than microscopic size be composed of cells. The capacity of cells to grow, in the sense of adding to their content of chemical substances that do not exist in the inorganic world, is limited. In general—though again there are exceptions—a cell can only double its mass. Thus any significant growth requires the production of new cells, and this involves the process of cell division. We can add continuously to the amount of living matter only by multiplication, and the multiplication takes place by division.

All this is commonplace today, but it seemed so unreasonable a century ago that it was not even accepted by Theodour Schwann and Matthias Jakob Schleiden, the two German biologists who are generally credited with the founding of the cell theory. Cell division had actually been seen and described many times by skillful microscopists. But an observation is seldom acceptable if its consequences seem unreasonable, and the dictum that every cell must arise by the division of a pre-existing cell did seem unreasonable. Schwann and Schleiden were much more reasonable when they proposed that new cells arise by a process analogous to the formation of crystals in a solution. They were merely wrong.

We now understand the reasons why cells must always arise from cells and by a process that seems excessively complicated. It is because the history and capabilities of a cell are determined by the threadlike chromosomes in its nucleus, and the various forms of life can maintain their historical continuity only if the chromosome patterns are transmitted from cell to cell. If the biological world is essentially conservative in its tendencies, it is because the chromosomes are so stable. The elaborate process of cell division that we call mitosis perpetuates them in all their stability.

Mitosis is the most complicated, and its machinery is about the most elaborate that a cell can manufacture. Hereinafter we shall refer to this machinery as the mitotic apparatus. Anyone who has ever attempted to make a satisfactory diagram of the process, or teach it to a class of normally unresponsive freshmen, will grant that its complexity is real. A machine of beautiful geometric perfection is temporarily created out of the seeming formlessness of the cell and operates with exquisite co-ordination. Let us consider what happens. We begin with the so-called "resting cell," a cell that is busy enough in other ways but has not yet gone through the travail of giving birth to two daughter cells. The chromosomes are contained in the nucleus and look like nothing at all. We know they are there from chemical and genetic data, but the lovely compact threads that we always visualise when we think of chromosomes usually are visible only during mitosis. In the resting cell there is no evidence of the mitotic apparatus, except that in favourable cases a small body appearing either single or double sometimes is detected near the surface of the nucleus. This is what we call the centriole or centrosphere.

What happens when this resting cell begins to divide may be regarded in two phases. First, the mitotic apparatus is formed. The centrioles divide, separate, and as they separate throw out what seem like radiating systems of fibres, appropriately called asters. These appear to find their way to opposite poles of the cell and to be connected by fibres running from one centriole to the other. This group of fibres is called the spindle. Meanwhile the previously invisible chromosomes are becoming visible as slender threads while the nucleus fades away. Then the threads become more compact. In a most remarkable way the chromosomes find their way to the equator of the spindle; that is, a plane exactly midway between the two centrioles and at right angles to a line connecting them. Here we have the culmination of this remarkable set of preliminary maneuvers. Now the mitotic apparatus is fully formed. At some point along the way it becomes possible to observe that each of the chromosomes has doubled. More precise evidence suggests that the real duplication process has occurred much earlier than the visible appearance of twin threads. In fact, it is now thought that the duplication takes place during the resting stage, which would mean that mitosis was a process for distributing the products of a reproductive step that had already been completed.

The second and perhaps more dynamic stage of mitosis is the movement of one complete set of chromosomes to each pole, and, co-ordinated with this, the division of the whole cell so that each set of daughter chromosomes goes to one of the daughter cells. It is the mitotic apparatus in action, moving the chromosomes apart, often pushing the cell out of shape, and somehow controlling the location of the wall that will finally separate the daughter cells. Finally we have two daughter cells, each with half of the mother's substance and each with a complete set of chromosomes seemingly identical with those of the mother. These daughters go back into a resting stage by a process that roughly resembles a movie of the first steps of mitosis run backward. The chromosomes disappear from view and a nucleus is formed.

The process has many variations. In plant cells we ordinarily do not see asters. Animal cells divide by what looks like a process of pinching in two; plant cells build a new wall. In both cases the barrier between the two daughter cells is formed exactly where the equator of the spindle was located. Some single-celled animals have bizarre systems of mitosis. In the bacteria the very existence of the mitotic system of cell division was doubted until recently. Edward DeLamater and Stuart Mudd of the University of Pennsylvania have reported that they can see in dividing bacteria what look like typical mitotic figures. Some students of microorganisms are not convinced; the difficulty rests on the sheer smallness of the cells and of the alleged mitotic figures. It is nonetheless significant that the trend is toward a unified picture of how cells divide.

The discovery of mitosis in the years around 1880 is a remarkable chapter in the history of biology, if only for its lack of drama. There was no one discoverer, although Walther Flemming of Germany may be credited with unifying the observations that he and others had made. There were no brilliant laboratory accidents here, but only powers of observation and imagination which awe the biologist of today. These men knew nothing about the significance of chromosomes; they had to observe the individual stages out of proper sequence and most often in dead cells. One wonders whether there are microscopists today who with this kind of information could piece together the logical sequence of mitosis in space and time.

Mitosis may be a rapid affair. It is not unusual for cells to go through the whole process in thirty minutes, and it is unusual for it to take more than a few hours. The preparatory phases take the longest; once the mitotic apparatus is fully formed, the actual process of division may take only a matter of minutes, almost as though the cells had snapped apart once the apparatus was there.

So much happens in this process that it would scarcely make sense to ask the question: "What is the mechanism of mitosis?" Knowing as much as we do of what takes place at the level of the microscope, and as little as we do of what happens below the level of the microscope, we will be wise to restrict our questions to the material nature of the mitotic apparatus, to the way in which it is formed and to the forces involved in the crucial movements of cell division.

The symmetry of the mitotic apparatus inspired a number of interesting physical models based on the operation of electrical and magnetic forces. In most cases it would appear that the chromosomes are being pulled through a liquid by special fibres which run from the poles and are attached to a definite region of each chromosome. This region has been called by various workers the centromere, kinetochore or simply "the spindle fibre attachment." The convincing reasons for this contractile fibre" hypothesis are two. First, each chromosome generally takes the form of the letter V or J or a rod, with the point facing the pole. This is what we would expect if a limp thread were dragged through a liquid. Second, the point of attachment is always the same for a given chromosome. If the point is eliminated when the chromosome is broken by X-rays or another agency, the chromosome does not move at all. There are, however, disturbing exceptions. Some years ago Charles W. Metz, now at the University of Pennsylvania, described a case in the fly *Sciara* in which the chromosomes pointed the wrong way! This has never been explained in terms of the contractile fibre hypothesis.

Various mechanisms have been proposed to explain the division of the cell mass, especially in those cases where a pinching or "furrow" is involved. One hypothesis involves the so-called "contractile ring." This hypothesis suggests that, at the time of division, a layer around the equator of the mitotic apparatus actually contracts and pinches the cell in two. It is based in part on some interesting experiments by Douglas Marsland of New York University. A contractile ring would have to be a rigid layer of colloid in the gel state. Such a gel, Marsland predicted, would become soft and liquid under high pressure, on the order of 500 atmospheres. Beneath the surface of cells there is a thin layer of gel called the cortex, and Marsland was able to show that the application of high pressure did make this layer more liquid.

Experiments on dividing cells revealed that pressures which liquefied the cortex also prevented the cell from dividing. Whether or not the contractile ring hypothesis proves to be true, these experiments demonstrate that a semirigid cortex plays an essential part in cell division.

A more detailed hypothesis on the mechanics of cell division in cases where asters are present and a furrow is formed has been elaborated by Katsuma Dan of Tokyo Metropolitan University in Japan. Some years ago Dan devised a very simple means of following the movements of the cell surface during division. He applied tiny particles of clay to the surface of cells, and some of these adhered. Then, with great patience, he measured the distance between two particles during cell division. His results were most unexpected. At the region of the furrow, where the surface would be expected to stretch, the distance between the particles actually decreased.

As the furrow forms, the cell elongates. The spindle, too, is longer from pole to pole. Dan pictures the spindle as a stiff body that pushes the centres apart. At this point the rays of the asters, growing in all directions, appear to have joined the surface of the cell. The rays of the opposite asters cross at the equator. Dan's ingenious proposal is that, when the asters are pushed apart, the tips of the rays are brought closer together in the region where they cross. If, as he imagines, the tips are attached to the cell surface, their movement will collapse it at a point midway between the centres.

Ordinarily the asters are of equal size, and the centre of the spindle is the equator of the cell; thus we expect the cell to divide into equal halves. But if one aster is much larger than the other the midpoint of the spindle will be displaced from the equator toward the smaller aster. In this case Dan predicts an unequal division, with the smaller daughter cell containing the smaller aster. This is confirmed by observation, and is a strong point in favour of his view.

In Dan's theory the work of cell division is accomplished by the push of the lengthening spindle at the time the cell is seen to divide. Last year two biologists at Cambridge University in England, M. M. Swann and J. M. Mitchison, proposed a mechanism which is a compromise between the contracting-ring theory, in which the mitotic apparatus has no role in the division of the cell mass, and Dan's theory, in which the role of the mitotic apparatus is paramount. According to Swann and Mitchison, the furrow is due to an active stretching of the cell surface by the unfolding of molecules that were previously rolled up and did not occupy as great an area. They consider that the mitotic apparatus may determine the position of the furrow and guide it inward as in Dan's model.

But neither through microscopic observation nor ingenious physical models can we achieve a feeling of intimate contact with the realities of cell division. The formulation has lacked substance, which in today's biology means a treatment in terms of identifiable molecules undergoing reactions translatable into the language of chemistry. What is the stuff of the mitotic apparatus and how are its constituents assembled into this beautiful little gadget? Answers to these questions solve no problems of mitosis, but put flesh between the bones of physical speculation and the clothing of microscopic appearance.

In the past year an opening toward the answers to these questions has been found in experiments conducted at the University of California by Dan and the author. The direct way to obtain the answers is clear enough. One should isolate the mitotic apparatus from dividing cells in pure form and in quantity and study its chemistry and its structure. Other bodies in the cell-nuclei, chromosomes, mitochondria—have been isolated and studied with great profit. But the problem here is a little more difficult. The mitotic apparatus is a temporary structure. It appears in the cell only when it is dividing, and changes continuously in the course of division. It is not freely suspended but is embedded in the substance of the cell. What is worse, it is extremely unstable in its living environment. If a dividing cell is damaged mechanically or treated with one of a number of chemicals, the mitotic apparatus simply disappears.

In order to experiment on the isolation of the mitotic apparatus in quantity, it was obviously necessary to work with a large population of cells, all of which were dividing synchronously. Dan and I were fortunate to have on hand, in the tide pools of the Pacific shore, the material of choice. For many years students of cell division have worked with the eggs of sea urchins, which may be induced to divide by fertilizing them with sperm cells. If a very large number of eggs are inseminated at a given time, all will divide synchronously. The whole mitotic process from the time of fertilization may take an hour or so, but so uniform is the cell population that 90 per cent of the cells will be seen to be forming division furrows within a period of five to ten minutes. Thus we could experiment with masses of cells which represented for the most part any desired stage of the mitotic process.

Because the mitotic apparatus is so unstable it was necessary to find a means of stopping the process with the least damage and the least chemical change. After many experiments, very cold (minus 10 degrees centigrade) dilute ethyl alcohol proved to be suitable.

Attempts to isolate the mitotic apparatus by the conventional method of mechanical disintegration of the cells were so discouraging that we turned to a fresh, though not necessarily original, approach. If the mitotic apparatus were chemically different from other things in the cell, could we not find some means of "dissolving" the cell in such a way as to liberate the apparatus? We knew of agents that dissolve (solubilise is the more precise word) cell structures without excessive damage to the molecules of which they are composed. Foremost among them are the detergents, substances with a soaplike action. At first we worked with synthetic detergents of the varieties commonly encountered in the dishpan and the laundry tub. We found that Duponol (sodium lauryl sulfate) was effective in solubilising the cytoplasm of the dividing sea-urchin eggs, but that it dissolved the mitotic apparatus as well.

Looking at the research in retrospect, it is evident that we should then have tested gentler detergents; it was perhaps fortunate that instead we attempted to analyse the conditions that determined whether or not a body would be solubilised by a detergent. Our approach may serve as an illustration of the role played by informed guesswork in research. First, we imagined that the mitotic apparatus was composed of proteins, an assumption for which there was slight factual basis. The cell generally employs proteins in the establishment of structure, and especially does so in the elaboration of fibrous structures. How protein fibres formed and what are are their properties? One mechanism of fibre formation is the establishment of the so-called disulfide (S-S) bridges between molecules. Protein molecules contain sulfur in the form of sulfhydryl (SH) groups. If, on two adjacent protein units, the hydrogens are removed (which is another way of saying that oxidation has occurred), the sulfur atoms of the two molecules may be linked in a disulfide bridge. Thus may fibrous chains be formed.

The disulfide bonds are very strong, so it is difficult to dissolve structures formed by them. In 1935 David Goddard and the late Leonor Michaelis, working at the Rockefeller Institute for Medical Research, showed that the reason why wool and similar proteins were so stable and insoluble in chemical solution was because they contained so many disulfide bridges. If, by the addition of an agent such as thioglycolic acid, the disulfide bonds were converted back to sulfhydryl bonds (a reversal of the oxidation described above), wool became soluble.

What does all this have to do with the mitotic apparatus? If the apparatus dissolves in Duponol or other strong detergents, the logical conclusion would be that the fibres of the spindle and asters are *not* formed by disulfide bridges. We drew the less logical conclusion that there were too few disulfide bridges to make the fibres sufficiently resistant. We then guessed—there is no other word that would be entirely honest—that there might be a larger number of *potential* disulfide bridges (SH groups) which could be converted to actual bridges (S-S bonds) by the addition of an oxidising agent. On this assumption we treated the eggs briefly with a solution of hydrogen peroxide before adding the detergent.

The experiment—every step of which was based on inference as to how the cell *might* make the fibres of the mitotic apparatus worked. The detergent dispersed everything in the cell except the mitotic apparatus, which was left intact! The contaminants could then be removed by letting the heavier mitotic apparatus settle out under a small centrifugal force. By repeated washing in this manner, a sediment composed of "pure" mitotic apparatus could be collected. The method set no limits on the number of dividing cells that could be used; it was possible to isolate enough of the mitotic apparatus for relatively large-scale chemical analysis.

The microscopic study of the isolated mitotic apparatus immediately supplies interesting information. Whatever the stage of mitosis, the whole apparatus may be isolated in one piece containing all the elements that entered into the description of mitosis. This proves that the cell does make a physically integrated apparatus for carrying out the cell-division process. An interesting result is obtained where one isolates the mitotic apparatus from clam eggs or other cells which do not divide into daughter cells of equal size. In these cases the asters are also of unequal size, the larger aster being at the end where the larger cell will form. Thus observations of isolated mitotic figures confirm Dan's theory as to how the plane of cell division is located. In general, observations of the isolated mitotic figures confirm the picture that had been obtained by earlier microscopists, but everything is much clearer and observation does not depend upon violent killing and staining procedures. The most interesting result of observing the isolated figures is the fact that they are isolated at all. This signifies that the dividing cell performs some chemical act, common to all parts of the apparatus, that makes it chemically different from everything else either in the dividing or the resting cell. This is a logical consequence of the fact that the isolation procedure works. While the procedure does not solve the "secret" of mitosis, it assures us at least that there is one, and one that can now be studied.

We now know something about the chemistry of the apparatus. It is composed mostly of protein, as was originally guessed. There is present a small amount of ribonucleic acid, and, of course, the contents of the chromosomes. Other constituents are suspected but not yet proved. The "weight" of a single mitotic apparatus has been determined: it is found that the apparatus involves about 2 per cent of all the protein in the egg at the time of its first division. This will certainly vary as cells of different kinds are investigated.

To what extent was the guess that the mitotic apparatus is formed by the linking together of protein molecules through sulfur bridges a legitimate one? A logical way to test this is by attempting to reverse the process. If the guesses were accurate—and the success of an experiment does not in itself prove that the assumptions which led to it were correct—then the mitotic apparatus should become soluble in detergent if the S-S bridges were reduced to unlinked SH groups. The experiment would be comparable to that by Goddard and Michaelis on the solubility of wool. The same reagent, thioglycolic acid, was used—and the predicted results were obtained. After treatment with this reagent, the mitotic apparatus was soluble in detergent. Therefore the idea that the fibres are formed through sulfur bridges is a tenable one. Once the protein of the mitotic apparatus is in solution, further information may be obtained. By the use of the ultracentrifuge, which measures how fast molecules settle under very strong gravitational forces, one can observe the size of the molecules. It appears that most of the apparatus consists of molecules about the size of "ordinary" proteins, such as the albumin of egg white. This might be considered the unit which the cell strings out by sulfur bridges to form the fibres of the mitotic apparatus.

While much could be learned from the mitotic apparatus isolated by peroxide and Duponol, the method was still a rough one. By the standards of the biochemist, peroxide is a rather violent reagent, and the isolation involved the creation of unnatural sulfur bridges which did not exist in the original cells. Following the same line of reasoning, now fortified by experiment, the problem could be simplified. If the "natural" mitotic apparatus depended upon disulfide bridges but contained relatively few, could we not use a detergent that was just strong enough to solubilise the other materials of the cell and just weak enough *not* to solubilise the mitotic apparatus? If such a detergent were available the peroxide could be omitted and the whole treatment would be gentler. After some trials such a detergent was found. It is a natural detergent: digitonin, which is obtained from the common foxglove. George Wald of Harvard University had used this substance some years ago to solubilise the visual pigments of the eye without damaging them chemically. Recent experiments in our laboratory have shown that digitonin alone will disperse the cell content and liberate the mitotic apparatus.

There is every reason to think that the isolated mitotic apparatus is now closer to its natural condition. Its structure remains intact down to fine details. A merit of the newer method is that the chromosomes, which were often difficult to see when isolated by the old method, are preserved. But again we must raise the question: was the reasoning that led to the successful isolation correct? If it was, the isolated mitotic apparatus could have fewer S-S groups and still be more soluble in thioglycolate than was the apparatus prepared by the old method. This proved to be the case.

Now a new and challenging fact emerges. Let us consider, for this soon becomes the heart of the problem, how the beautiful geometry of the mitotic apparatus might be achieved. On visual evidence alone the old masters of microscopy believed that the centrioles were in fact centres which formed the fibres of the mitotic apparatus. Since the appearance of these bodies varies so much from cell to cell, let us use the general term "centre." The fibres do seem to grow out of the centres. No alternative hypothesis has been proposed, although it is often suggested that one group of fibres, those connecting the chromosomes to the poles, may originate in the chromosomes. Translating into chemical language, we may set up the working hypothesis that the form of the mitotic apparatus is determined by centres at which fibres are made by a process involving the formation of disulfide bonds. The peculiarity of these centres would be that they provide conditions favouring such a polymerisation. Recent observations on the mitotic apparatus isolated by the digitonin method lend some support to such a line of thought. If the mitotic apparatus is given a very mild exposure to thioglycolic acid, everything dissolves but the spherical central region of the asters. This seems to be in fact the region where the S-S bridging of the proteins is strongest. We are a long way from understanding how the mitotic apparatus is formed, but the availability of the isolated apparatus and now the isolated centres should permit us to feed chemical information into the theory of mitosis as it evolves.

Mitosis cannot be understood merely as a series of chemical transformations. It is a dynamic mechanical process, a matter of push and pull, stress and strain. What tools have we to study this aspect of the problem? The most refined tool, and one that is being used most effectively, is the so-called polarisation microscope. Where molecules are oriented, and take on other than random arrangements, they influence the speed and direction of beams of polarised light passing through them. The orientation may be the result of the architectural operations of the cell, as in the formation of the mitotic apparatus, or the result of mechanical stresses. The influence of an object composed of oriented units on a beam of polarised light is measured as its birefringence. Birefringence is actually the difference between the speed of a beam of polarised light through the object when the light waves are vibrating parallel to an axis of the object, and the speed of a beam vibrating at right angles to this axis. The axes are obvious in the case of an object such as an elongated fibre. What is important is that the birefringence may measure the degree of orderliness in the arrangement of the molecules or the extent to which the molecules themselves are nonsymmetrical. In the case of a fibre composed of long and slender units, the birefringence will increase if all the molecules are made to lie parallel to one another. It will decrease if for any reason the individual molecules become shorter or fatter.

Researchers whose main interest has been the problem of mitosis have in recent years refined the polarisation microscope to the point where we may now hope for a real surge of information about the arrangement of molecules in the mitotic apparatus. A few examples will illustrate the kinds of information that we can expect. Some years ago a German pioneer in the field, W. J. Schmidt, showed that the mitotic spindle was birefringent, and that the birefringence decreased when the chromosomes were separating. This demonstrated that the fibre system of the spindle had as its basis a system of fibres of molecular dimensions, aligned more or less parallel to the visible fibres. The decrease in birefringence indicated either that fewer of the molecules were aligned along the axis of the spindle or that the molecules were contracting. Schmidt's methods were not sufficiently sensitive to detect smallscale changes. Recently Swann and Mitchison have refined the microscope so that much smaller degrees of birefringence can be measured, and the measurement can be made on a movie of the dividing sea-urchin egg. They were able to show that the decrease in birefringence as the chromosomes moved apart was not uniform, but began near the chromosomes and spread toward the poles. Swann has proposed that the changes that take place in the spindle during chromosome separation are initiated by one or more substances produced by the chromosomes. An engaging feature of this hypothesis is the way in which it demonstrates how a few facts (even when incorporated into a hypothesis that is difficult to prove) may resolve a seeming contradiction. For years cytologists have debated whether the chromosomes are pulled toward the poles or themselves provide the momentum for their migration. Swann's experiment reveals that both the spindle and the chromosomes may play an active part. This is often the outcome of such "either-or" controversies.

Microscopic research on mitosis has been both stimulated and plagued by debates as to whether this or that part of the mitotic apparatus was real or was an "artifact" produced by the technique of preparing the cell for observation. At one time the reality of even the chromosomes was debated; the question of their existence was only resolved by the ultimate court of appeal—observation in the living state. The controversy as to the reality of the fibres in the mitotic apparatus has continued to rage, though observations of the fibres in living cells have been reported by Kenneth Cooper of the University of Rochester and by others. In the opinion of the writer, the polarisation microscope has solved this problem. Shinya Inoué, a young graduate student working under Cooper at Princeton University, designed and built a polarisation microscope capable of detecting and measuring smaller degrees of birefringence than had previously been possible in biological material. With this instrument he has studied the mitotic figures in many kinds of living cells, and in every case he has been able to detect distinct spindle fibres. The fibres, which contrast so poorly with the background material in ordinary light, stand out vividly in polarised light by virtue of the arrangement of the molecules in them.

The reader is sure to ask about the contributions of the electron microscope to the study of mitosis. Living material cannot be examined in this instrument because the beam of electrons with which it resolves very small objects must be focused in a vacuum. Thus the application of the electron microscope to the study of mitosis is already beset by arguments as to whether the observations made with it are real or artifacts.

The difficulties could have been predicted from the troubles of pre-electron microscopists. "Seeing" depends on contrast, and the electron microscope, however great its magnifying power, depends on a rather high degree of contrast (difference in density of matter between an object and its surroundings) in order to show an object at all. It is perhaps for this reason that the mitotic apparatus, observed while it was embedded in the other constituents of the cell, has seemed so structureless even to skilled electron microscopists. On the other hand, the isolated mitotic apparatus promises to show a great deal of structure. Patricia Harris of our laboratory has made photographs of the isolated mitotic apparatus which, even at the relatively low powers employed thus far, reveal a number of interesting details. It seems, for example, that the spindle fibres are not straight and continuous from pole to pole, but appear to be formed by an end-to-end union of fibres from either pole at the equator. As everyone would have predicted, the fibrous appearance of the spindle is not due to distinct fibres with nothing between them; the spindle is more like a sheaf of threads with a tendency to lie side by side. What will be seen as the full resolving power of the electron microscope is brought into play cannot yet be predicted. So powerful is this tool that we are forced to proceed slowly toward the highest magnifications, using each step to provide the landmarks for the next.

If we survey the whole living world, we see that mitosis is the most common mechanism for distributing the hereditary birthright, and the most fundamental features of mitosis are the formation of a "plate" of doubled chromosomes and the formation of a spindle which will guide each of the daughters of chromosome doubling to the opposite pole. There are important exceptions to this typical pattern. In the growth of our own muscles, for example, nuclei may reproduce by mitosis without the formation of new cells around them. Growth keeps pace with the number of nuclei, and the result is huge cells with many nuclei. There are instances where, after the division of nuclei alone has given rise to a large cell, the cell suddenly builds a wall around each nucleus so that we end up, after all, with as many cells as there are nuclei. This process is of great practical importance, for the formation of infectious spores in many microorganisms is due to it. How new cells can be marked out without a mitotic apparatus to guide the process we do not understand at all.

As mentioned earlier, we cannot detect asters in many cells, especially those of plants. Nor have the centres toward which the chromosomes of plants migrate been identified visually as the distinct bodies called centrioles. In some very common instances, such as the division of white blood cells and other cells that crawl like amoebae, the division is accompanied by lively movements. Such divisions look for all the world as though, once the forming daughter cells decide that they are daughter cells, they try to crawl away from their sisters and thus aid the division process. In many but not all of the protozoa, the one-celled animals, there is a peculiar kind of mitosis in which the nuclear surface does not break down. The spindle forms inside the nucleus, which elongates as the daughter chromosomes part, and finally pinches in two. Even here, where we can see neither asters nor mitotic apparatus outside the nucleus, the cell "knows" where the equator of the peculiar spindle is, for it pinches into two at the midpoint. We also have, in such diverse instances as mosquito gut cells and certain plant cells, "endomitosis," which involves neither cell division nor even the separation of the daughter chromosomes into new nuclei. The chromosomes simply divide and remain together in the same nucleus.

The advance of biology is an interplay of generalisation and exception, and attitudes toward exceptions tell us more about biologists than they do about biology. To the writer it seems that the diverse patterns of cell multiplication will emerge as variations on a simple design once we learn something of what lies beneath the surface.

In the full context of nature the biological world is characterised by continuous growth, limited only by the resources and hazards of the physical world. But, just as human beings fancy themselves as being more than mere descendants and ancestors, so the life of an organism is largely a vacation from the propagation of its kind. Even single cells do not add to their store of living matter during division, but only between divisions. In higher organisms the pattern of development is a counterpoint of division and organisation. In the embryo the frequency of mitosis decreases before the forms and specialisations of the adult come into being. When a group of cells reverses this pattern, we have a cancer. What sets mitosis into motion and what stops it? We have only a few scattered facts to marshal against these questions. In general, the mass of a cell must reach a critical maximum value before it begins to divide. If we keep amputating part of an amoeba so that it never reaches full size, it never divides. We also have learned in the last few years that mitosis is a process which distributes equally between the daughter cells chromosomes which have doubled before mitosis begins. It is conceivable that the control of mitosis involves restraints on the synthesis of new chromosome material.

What artificial restraints can we place on mitosis, short of simply murdering cells? We know several kinds. One is to suppress the synthesis of the nucleic acids of the chromosomes. This is what happens when we expose cells to mild doses of X-rays or other ironising radiation from which they generally recover. Heavier doses of radiation upset mitosis by breaking up the chromosomes. So far as we know, most biological effects of radiation involve the formation and the integrity of the chromosomes. Chemicals that duplicate these effects are called radiomimetic—mimickers of radiation. On the other hand, ionising radiation seems to have little effect on the mitotic apparatus apart from the chromosomes, but many chemicals do. Substances such as the plant derivative colchicine prevent the formation of the mitotic spindle, but do not affect the chromosomes. The chromosomes go through their complete mitotic cycle, but cannot be separated because there is no mitotic apparatus. The result is a single cell with a double portion of chromosomes. Such antimitotic agents have been used successfully in the palliation, but not the cure, of cancer, and have been useful where it was desirable to breed organisms such as food plants with twice the parental ration of chromosomes. It goes without saying that a reasonable approach to the problem of controlling mitosis will depend on understanding mitosis itself.

Biology has few generalisations that can be dignified by the word "law," and to most of these there are exceptions. But no exception has been discovered to the law that every cell must arise from another cell. This generalisation makes sense once we have discovered the essential accomplishment of mitosis: the deployment of the chromosomes so that each daughter cell is assured a full representation of all the chromosomes in the mother cell.

The essential meaning of mitosis can be learned from microscopic observation alone, and was learned many years ago. Now we begin to see some progress toward a deeper understanding of what actually is happening in the dividing cell. Through the combined efforts of those who study virus reproduction and those who investigate cell division, we may hope to have some insight into the mechanism whereby the chromosome material is duplicated before mitosis begins. But there are aspects of the mitotic process that do not lend themselves so well to analysis by current means. Should we ask how the chromosomes come to align themselves in the correct geometric plane before they separate and how sister chromosomes find their way to opposite poles, we discover that we are dealing with an order of problems that does not fit into current methods or ways of thinking about biological events. While we are learning a great deal about the chemistry of cells, we are a long way from understanding how any part of the cell knows where it is.

Cell Culture

HEK-293 (human embryonic kidney) cells stably transfected with FLAG (Asp-Tyr-Lys-Asp-Asp-Asp-Asp-Lys)-tagged human 11 [β]-HSD2 were grown at 37°C under 5% carbon dioxide to 60-70% confluence in Dulbecco's modified Eagle medium (DMEM) supplemented with 10% fetal calf serum, 4.5 g/L glucose, 50 U/mL penicillin/streptomycin, and 2 mM glutamine. SW-620 (human colourectal adenocarcinoma) cells and JEG-3 (human choriocarcinoma) cells were cultured according to the recommendations of the supplier. The recently described RCCD-2 aldosterone-sensitive rat cortical collecting duct cells were cultured in DMEM/Ham's F-12 (1:1), 14 mM NaHC[O_3], 2 mM glutamine, 10 U/mL penicillin/streptomycin, and 20 mM HEPES, pH 7.4.

Activity assays in cell lysates. To measure 11[β]-HSD2 activity, stably transfected HEK-293 cells were grown in 10 cm culture dishes to 90% confluence. Cells were rinsed 3 times with phosphate-buffered saline (PBS) and resuspended in 2 mL ice-cold buffer TS2 (100 mM NaCl, 1 mM EGTA, 1 mM EDTA, 1 mM Mg[Cl_2], 250 mM sucrose, 20 mM Tris-HCl, pH 7.4). Aliquots of the cell suspension were frozen at -20°C, retaining full enzymatic activity for at least 3 months. For determination of 11[β]-HSD2 activity, aliquots were thawed, sonicated, and diluted 1:12 in buffer TS2 (4°C). We carried out reactions in a final volume of 20 [micro]L containing 10 nCi [1,2,6,7-[^{3}]H]-cortisol, 400-[micro]M [NAD^+], and different concentrations of unlabelled cortisol. Final cortisol concentrations were 40 nM for measurements of inhibitors and ranged between 10 nM and 200 nM for determination of apparent [K_m] values. Incubations were for 10 min at 37°C.

For determination of 11[β]-HSD1, 17[β]-HSD1, and 17[β]-HSD2 activity, HEK-293 cells transfected by the calcium-phosphate precipitation method were harvested 48 hr later, washed with PBS, and centrifuged for 3 min at 150 x g. Supernatants were removed, cell pellets quick-frozen in a dry ice ethanol bath, and stored at -70°C. For assaying 11[β]-HSD1, we dissolved pellets in TS2 buffer; for 17[β]-HSD1 or 17[β]-HSD2, we used a buffer containing 50 mM potassium phosphate, 20% glycerol, and 1 mM EDTA. We measured 11[β]-HSD 1 oxoreductase activity as described recently, using radiolabelled cortisone as substrate. 17[β]-HSD1 and 17[β]-HSD2 activities were measured in the presence of radiolabelled estrone or estradiol and unlabelled steroid at final concentrations of 200 nM and 500 [micro]M NADPH or [NAD^+], respectively.

Determination of 11[β]-HSD2 activity in intact cells and MTT cytotoxicity assay. HEK-293 cells stably transfected with 11[β]-HSD2 (25,000 cells per well) were seeded 24 hr prior to the assay in poly-D-lysine coated 96-well Biocoat plates (Becton Dickinson, Basel, Switzerland). The medium was carefully removed, followed by the addition of 30 [micro]L fresh medium, 10 [micro]L medium containing various concentrations of organotins, and 10 [micro]L radiolabelled cortisol. The reaction volume was 50 [micro]L, with a final cortisol concentration of 40 nM. The cells were incubated for 2 hr at 37° C under 5% C[O_2]. We stopped reactions by adding an excess of unlabelled cortisone and cortisol in methanol, and separated steroids using thin-layer chromatography, followed by scintillation counting. 11[β]-HSD2 activity in JEG-3, SW-620, and RCCD-2 cells was measured similarly by adjusting the cell density and reaction time to obtain a maximal conversion of cortisol between 15 and 25%. To ensure that the observed inhibition of 11[β]-HSD2 activity was not due to cell death, we assessed cytotoxicity of organotin compounds parallel to the activity assay under identical conditions. The corresponding organotin compound was added to the cells, followed by incubation for 2 hr at 37° C under 5% C[O_2]. Cells were washed with PBS and incubated in fresh medium containing 0.5 mg/mL 3-(4,5-dimethylthiazol-2-yl)2,5 diphenyltetrazolium bromide (MTT). After conversion of MTT, we removed the medium and added 200 [micro]L DMSO to the insoluble fraction. Conversion of MTT was kept below OD 0.9.

Site-Directed Mutagenesis and Analysis of Mutant 11[β]-HSD2 Enzymes

Mutations were introduced into the C-terminally FLAG-tagged 11[β]-HSD2 cDNA in Bluescript vector by site-directed mutagenesis according to the Quick Change mutagenesis kit (Stratagene,

Amsterdam, the Netherlands). All constructs were verified by sequencing and recloned into pcDNA3 expression vector. Wild-type and mutant enzymes were expressed in HEK-293 cells, lysates were prepared, and proteins were separated by sodium dodecyl sulfate (SDS) gel electrophoresis. Proteins were transferred to nitrocellulose, and expression levels of 11[β]-HSD2 constructs were detected using mouse monoclonal antibody M2 raised against the FLAG epitope and visualised with a horseradish peroxidase conjugated anti-mouse antibody and enhanced chemiluminescence Western kit. After detection of 11[β]-HSD2 constructs, nitrocellulose membranes were stripped and incubated with rabbit polyclonal anti-actin IgG and horseradish peroxidase-conjugated goat anti-rabbit IgG to adjust for the amount of proteins loaded on the gel. The expression of mutant relative to wild-type enzyme was adjusted for calculation of kinetic parameters.

Three-Dimensional Modelling

The 3-D model of human 11[β]-HSD2 from Arnold et al. (2003) was used with [NAD^+] extracted from protein data bank file 1AHI. Human 11[β]-HSD2 was then minimised for 100,000 iterations with Discover 3 using the extensible and systematic force field (ESFF), with a distant dependent dielectric constant of 2, to model water in the protein.

Statistical Analysis

Enzyme kinetics were analysed by nonlinear regression using Data Analysis Toolbox (MDL Information Systems Inc., Nashville, TN, USA) assuming first-order rate kinetics. Data represent mean [+ or -] SD of at least four independent experiments.

Inhibition of 11[β]-HSD2 but not 11[β]-HSD1 by Organotins

To investigate whether organotins disrupt the control of the ratio of active to inactive glucocorticoids by inhibition of 11[β]-HSD enzymes, we incubated lysates of HEK-293 cells expressing recombinant 11[β]-HSD constructs with various concentrations of TBT and TPT and determined enzymatic activities. TBT and TPT did not inhibit the 11[β]-HSD1-dependent conversion of cortisone to cortisol at concentrations up to 200 [micro]M. In contrast the 11[β]-HSD2-dependent conversion of cortisol to cortisone was efficiently inhibited by both compounds with [IC_{50}] (median inhibitory concentration) values in the low micromolar range, indicating that organotins selectively abolish the 11[β]-HSD2-dependent inactivation of glucocorticoids.

Despite that 11[β]-HSD1 and 11[β]-HSD2 interconvert the same substrate, they are phylogenetically relatively distant enzymes, sharing only 18% identical amino acid sequence. 11[β]-HSD2 is more closely related to 17[β]-HSD2, with an amino acid sequence about 45% identical. Thus, we also determined the conversion of estrone to the more potent estrogen estradiol by 17[β]-HSD1 and the reverse reaction by 17[β]-HSD2 in the presence of various concentrations of organotins. Neither TBT nor TPT inhibited 17[β]-HSD1. 17[β]-HSD2 activity was inhibited by TPT at a 6-fold higher [IC_{50}] value compared with that of 11[β]-HSD2, indicating that organotins preferentially inhibit 11[β]-HSD2.

Because trialkyltins are progressively dealkylated by microorganisms in the environment and in mammalian organs including brain, liver, and kidneys, and because the dialkyltins DBT and DPT are major metabolites with significant toxicity, we included these two compounds in the study. Analysis of all four organotin compounds revealed comparable inhibitory properties both in lysates and intact cells expressing 11[β]-HSD2 with [IC_{50}] values in the low micromolar range. In intact cells the trialkyltins were approximately 3-fold more potent than the dialkyltins, and the phenyltins were slightly more potent than the butyltins, well correlating with the hydrophobicity of these compounds. It is important to note that the inhibitory potency of organotins is comparable with that of the well-known 11[β]-HSD inhibitor GA in intact cells.

Dithiothreitol but not glutathione protects 11[β]-HSD2 from inhibition by organotins. To investigate the molecular mechanism of organotin-induced inhibition of 11[β]-HSD2, we measured the effect of organotins in the presence or absence of the reducing agent dithiothreitol (DTT). DTT alone did not significantly alter enzymatic activity. A concentration of 2 mM DTT, added simultaneously with 5 [micro]M of the corresponding organotin compound, restored 70-80% of 11[β]-HSD2 activity. Upon preincubation of 11[β]-HSD2 with organotins for 15 min, 50-60% of the enzymatic activity could be restored (not shown), indicating that most but not all of the inhibitory effect was reversible. In contrast to the dithiol DTT, the monothiol glutathione was not able to prevent organotin-dependent inhibition of 11[β]-HSD2.

We recently demonstrated that dithiocarbamates, another environmentally relevant class of compounds, inhibit 11[β]-HSD2. Cofactor [NAD^+] partially protected the enzyme, suggesting that covalent modification of the thiol group at [Cys^{90}] in the cofactor-binding region may be responsible for dithiocarbamate-induced inactivation of 11[β]-HSD2. In contrast we found no protective effect of [NAD^{+}] on organotin-induced enzyme inhibition, indicating an inhibitory mechanism distinct from that of dithiocarbamates.

Functional Analysis of the Cysteine Residues of 11[β]-HSD2

We have previously shown that substitution of [Cys^{90}] by serine leads to abolished protein expression and enzymatic activity. Because [NAD^{+}] did not protect 11[β]-HSD2 from organotin-induced inhibition, we analysed the functional relevance of the remaining eight cysteine residues by mutating them to serines. All nine cysteineto-serine mutants contained a FLAG epitope at the C-terminus, allowing the quantification of the relative protein expression. Five of the mutant enzymes showed significantly reduced expression. No band could be detected for mutant [Cys^{228}]Ser and only a very weak band for [Cys^{90}]Ser was detected. As expected, no activity could be measured using lysates from cells transfected with cDNA for either of these two mutant enzymes. Significantly reduced expression was also observed for mutants [Cys^{128}]Ser and [Cys^{188}]Ser. When the catalytic activity of these mutant enzymes was adjusted for their reduced expression, an apparent [V_{max}] comparable to that of wild-type 11[β]-HSD2 was obtained, indicating reduced enzyme stability but intact catalytic activity. For mutant [Cys^{264}]Ser, which also showed reduced expression, a slightly higher [V_{max}] Was obtained. Comparison of the kinetic parameters revealed altered kinetic parameters for mutant [Cys^{264}]Ser, with a 3-fold higher apparent [K_m]. Mutation of [Cys^{264}] to serine also resulted in a 2-fold increase in the [IC_{50}] for TBT, suggesting a role for [Cys^{264}] in the binding of cortisol and in the interaction with organotins with 11[β]-HSD2.

Analysis of the Mode of Organotin-Dependent Inhibition of 11[β]-HSD2

We next investigated the effect of preincubation of 11[β]-HSD2 with either 1.5 [micro]M TPT or 2 [micro]M TBT for 5 or 10 min. Although there was a slight tendency toward increased inhibitory effect upon preincubation with TBT, the changes did not reach significance, in line with reversible inhibition. To further assess the mode of inhibition, we determined the change of apparent [K_m] and [V_{max}] in the absence or presence of TBT. The apparent [K_m] increased 2- and 3-fold upon incubation with 2 [micro]M and 3 [micro]M TBT

(62 [+ or -] 17 nM for the control compared with 137 [+ or -] 24 nM and 197 [+ or -] 33 nM for treated samples, respectively), whereas [V_{max}] decreased slightly (2.14 [+ or -] 0.32 nmol x [h^{-1}] x [mg^{-1}] for the control compared with 1.66 [+ or -] 0.40 nmol x [h^{-1}] x [mg^{-1}] and 1.48 [+ or -] 0.48 nmol x [h^{-1}] x [mg^{-1}] for treated samples, respectively). These findings suggest a mixed-competitive mode of inhibition, with most of the inhibitory effect being reversible.

To further test this assumption, we measured the effect of diluting the enzyme-inhibitor complex (EI). In case of a competitive mode of inhibition, dilution of the EI complex would lead to a decreased inhibitor concentration and reduced competition with the substrate, hence reduced enzyme inhibition. If covalent enzyme modification takes place, the dilution of the El complex would not change the proportion of modified to unmodified molecules, meaning that after the dilution of El there would be no change of the relative inhibition compared with the control. In case of TBT-induced inhibition of 11[β]-HSD2, a 2-fold and 4-fold dilution of the El complex led to a proportional decrease of the relative inhibitory effect, indicating a transient interaction of the organotin compound with 11[β]-HSD2.

Chapter 4

Sulphotransferase and Polychlorobiphenylols

Sulphonation is a major phase II biotransformation reaction. In this study, we found that several polychlorobiphenylols (OH-PCBs) inhibited the sulphonation of 3-hydroxybenzo[a]pyrene (3-OH-BaP) by human liver cytosol and some cDNA-expressed sulphotransferases. At concentrations > 0.15 [micro]M, 3-OH-BaP inhibited its own sulphonation in cytosol fractions that were genotyped for SULT1A1 variants, as well as with expressed SULT1A1*1, SULT1A1*2, and SULT1E1, but not with SULT1A3 or SULT1B1. The inhibition fit a two-substrate kinetic model. We examined the effects of OH-PCBs on the sulphonation of 0.1 or 1.0 [micro]M 3-OH-BaP, noninhibitory and inhibitory substrate concentrations, respectively. At the lower 3-OH-BaP concentration, OH-PCBs with a 3-chloro-4-hydroxy substitution pattern were more potent inhibitors of cytosolic sulfotransferase activity [with concentrations that produced 50% inhibition (I[C_{50}]) between 0.33 and 1.1 [micro]M] than were OH-PCBs with a 3,5-dichloro-4-hydroxy substitution pattern, which had I[C_{50}] values from 1.3 to 6.7 [micro]M. We found similar results with expressed SULT1A1*1 and SULT1A1*2. The OH-PCBs were considerably less potent inhibitors when assay tubes contained 1.0 [micro]M 3-OH-BaP.

The inhibition mechanism was noncompetitive, and our results suggested that the OH-PCBs competed with 3-OH-BaP at an inhibitory site on the enzyme. The OH-PCBs tested inhibited sulphonation of 3-OH-BaP by SULT1E1, but the order of inhibitory potency was different than for SULT1A1. SULT1E1 inhibitory potency correlated with the dihedral angle of the OH-PCBs. The OH-PCBs tested were generally poor inhibitors of SULT1A3- and SULT1B1-dependent activity with 3-OH-BaP. These findings demonstrate an interaction between potentially toxic hydroxylated metabolites of PCBs and polycyclic aromatic hydrocarbons, which could result in reduced clearance by sulphonation. Polycyclic aromatic hydrocarbons (PAHs) and polychlorinated biphenyls (PCBs) are two classes of environmentally prevalent pollutants. PAHs are formed through the combustion of fossil fuels and the burning of organic materials. PCBs were first produced industrially in the middle of the last century for their desirable dielectric properties and remain in the environment because of their continued use, because of their release from waste sites, and because many congeners are slowly degraded. The more lipophilic PAHs and PCBs are often found in the same environmental samples, such as soils and sediments, and are biotransformed in animals by similar pathways.

Of the PAHs, benzo[a]pyrene (BaP) is a well-studied chemical carcinogen, which is metabolised by cytochrome P-450 (CYP) to a variety of products. These include 3-hydroxybenzo[a]pyrene (3-OH-BaP), a major metabolite of BaP in humans and animals, which has estrogenic properties and binds to hemoglobin. Hydroxylated PAH metabolites such as 3-OH-BaP are substrates for glucuronidation and sulphonation, catalysed by one or more of the UDP-glucuronosy-ltransferases and 3'-phosphoadenosine 5'-phosphosulphate (PAPS)-dependent sulfotransferases (SULTs), respectively. Sulphonation is considered a detoxification pathway for 3-OH-BaP.

PCBs have several metabolites of toxicologic importance, including the polychlorobiphenylols (OH-PCBs), which are formed in vivo from CYP-dependent mono-oxygenation of PCBs. Although they are slightly more hydrophilic than are the parent PCBs, several OH-PCBs are eliminated slowly. People who are highly exposed to PCBs through the diet typically have OH-PCBs in their blood, some bound to plasma proteins. Several OH-PCB congeners interact with components of the endocrine system, potentially interfering with thyroid hormone and estrogen function. Although the OH-PCBs have low affinities for both [α] and [β] estrogen receptors, some OH-PCBs are strikingly potent inhibitors of human estrogen sulfotransferase (SULT1E1), with subnanomolar concentrations that produced 50% inhibition (IC50). This suggests that OH-PCBs may be indirectly estrogenic by increasing estradiol bioavailability in target tissues. As well as possibly causing toxicity by inhibiting the sulphonation of hormones, several OH-PCBs inhibited the sulphonation and glucuronidation of the PAH metabolite 3-OH-BaP in channel catfish intestine.

Sulphonation is an important phase II conjugation pathway for the detoxification of xenobiotics as well as the modulation of endogenous compounds such as thyroid hormones, steroids, and neurotransmitters. One or more members of a superfamily of cytosolic SULT enzymes catalyse these reactions. SULT1A1, SULT1B1, and SULT1E1 are the major phenol sulfotransferases expressed in human liver, with SULT1A1 (also known as ST1A3) found at the highest concentration. SULT1A3 is expressed in the gut but is present in very low concentrations in adult human liver. Genetic polymorphisms are known for SULT1A: a [G^{638}] [→] A transition leading to an [Arg^{213}] [→] His exchange in the protein was observed with a frequency of 33.2% in Caucasian subjects, 8% in Chinese, and 29.4% in African Americans. SULT1A1*His (SULT1A2*2) was a less thermostable protein than SULT1A1*Arg (SULT1A1*1), and some authors have reported that the SULT1A1*2 variant is less catalytically active.

Because people are frequently coexposed to PAHs and PCBs, we wished to determine if OH-PCBs would inhibit 3-OH-BaP sulphonation in human liver (HL) cytosol and, if so, whether the inhibition was isozyme selective. We used cDNA-expressed human SULT1A1*1, -1A1*2, -1A3, -1B1, and 1E1 isozymes, which we expected would use 3-OH-BaP as substrate. We genotyped the HL cytosol fractions used in this study, with respect to the common SULT1A1 polymorphic variants, to examine the possibility that OH-PCBs would affect their activity differently. These studies were conducted with a series of predominantly para-OH-PCBs.

In naming these OH-PCBs, we followed the recommendation of Maervoet et al. (2004) to name them as metabolites of PCBs, referring back to the Ballschmiter and Zell numbering system for PCBs. The 6'-OH-CB35 (A1), 4'-OH-CB35 (B1), 4'-OH-CB36 (B2), 4'-OH-CB79 (C1), and 4-OH-CB36 (C2) were synthesized by Suzuki coupling as described previously. We verified the structures of each of these OH-PCBs by ^{1}H and ^{13}C-nuclear magnetic resonance spectroscopy, Fourier transform infrared spectroscopy, and gas chromatography-mass spectrometry (GC-MS). We found that each OH-PCB was > 99% pure by GC-MS analysis, combustion analysis, and thin-layer chromatography. The 4'-OH-CB69 (B3), 4'-OH-CB106 (B4), 4'-OH-CB112 (B5), 4'-OH-CB121 (C3), 4'-OH-CB159 (C4), 4'-OH-CB165 (C5), and 4'-OH-CB72 (C6) were purchased from AccuStandard. S.S. Singer (University of Dayton, Dayton, OH) supplied the PAPS. We purchased [^{35}]S-PAPS, 3.05 [micro]Ci/nmol (99.1% pure), from PerkinElmer Life

Science. Benzo[a]pyrene-3-sulphate (BaP-3-S[O_4]) and 3-hydroxybenzo[a]pyrene (3-OH-BaP) were purchased from the NCI Chemical Carcinogen Reference Standard Repository (Midwest Research Institute, Kansas City, MO). We obtained HaeII from Fisher Scientific (Atlanta, GA) and Taq DNA polymerase, along with other polymerase chain reaction (PCR) reagents, from Promega. Integrated DNA Technologies (Coralville, IA) supplied primers for use in genotyping. We purchased the highest available grade of other reagents from Fisher Scientific and Sigma Chemical Company (St. Louis, MO).

Physicochemical Properties of the OH-PCBs

We calculated the structural characteristics of dihedral angle, molecular volume, molecular surface area, p[K_a], log P, and log D at pH 7.0 with MM[2*] using GB/SA water solvent continuum as implemented by MacroModel 5.0 and described previously by Tampal et al. (2002).

Cytosolic Preparations

F.P. Guengerich (Vanderbilt University) kindly donated the samples of human liver, which were procured from organ donors. We prepared liver cytosolic fractions from four livers by standard methods and stored aliquots at -80°C until use. We used three or four of these cytosol fractions in each experiment.

SULT1A1 Genotype Determination

We used a genomic DNA isolation kit to extract genomic DNA from samples of the individual human livers used in this study. We used a published method to detect the SULT1A1 polymorphism status of each liver. Amplified DNA fragments were digested with HaeII, and the fragments were resolved on 2% (weight/volume) agarose gels. Fragments from individuals homozygous for SULT1A1*1 exhibited two bands, visualised by ultraviolet transillumination, whereas DNA fragments from individuals homozygous for SULT1A1*2 were not cleaved by this enzyme and exhibited one band.

Expression and Partial Purification of SULT Isozymes

The expression of human SULT1A1*1, SULT1A3, SULT1B1, and SULT1E1 in Escherichia coli has been described previously. We grew E. coli cells containing the respective sulfotransferase genes as described previously, and prepared 100,000 g supernatant fractions for use in inhibition studies and for partial purification of the SULT enzymes. We purchased expressed SULT1A1*2 cytosolic extract from PanVera and used it as supplied. The 100,000 g supernatant fractions of SULT1A1*1, SULT1A3, SULT1B1, and SULT1E1 were partially purified by chromatographic methods. After the last step, a 3'-phosphoadenosine 5'-phosphate (PAP)-agarose affinity column, we dialysed the fractions eluted with PAP with three changes of buffer to remove PAP before the assay of SULT activity with 3-OH-BaP as substrate. We analysed active fractions by SDS-PAGE (Laemmli 1970) to assess the purity of each SULT enzyme. We stained the gels with Coomassie R-250 reagent and determined the percentage of protein present as each respective SULT enzyme by scanning densitometry.

Kinetic Analysis of 3-OH-BaP Sulphonation

We determined SULT activity with 3-OHBaP as substrate by a fluorimetric assay of BaP-3-S[O_4] product formation, as described previously. We ensured that the formation of BaP-3-S[O_4] was linear for time and protein and did not exceed 10% of the added 3-OH-BaP with each of the enzyme sources used. Duplicate tubes were prepared for each incubation condition. We examined the kinetics of sulphonation in three liver cytosol fractions by systematically varying the concentration of 3-OH-BaP or PAPS. When the variable substrate was 3-OH-BaP, we used 12 concentrations in the range from 0.035 to 2.00 [micro]M, and the concentration of PAPS was kept constant at 10 [micro]M. When we varied PAPS, we used 7 concentrations from 0.157 to 10.0 [micro]M and kept the concentration of 3-OH-BaP constant at 0.100 [micro]M. We determined the kinetic parameters for 3-OH-BaP sulphonation by partially purified preparations of the cDNA-expressed SULT isozymes under incubation conditions similar to those used for liver cytosol. For SULT1A1*1 and -1A1*2, we used seven substrate concentrations in the range from 5 to 100 nM; for SULT1E1 we used six 3-OH-BaP concentrations from 15.6 to 1,000 nM; and for SULT1A3 and 1BI we used seven concentrations of 3-OH-BaP from 0.25 to 5.0 [micro]M.

Inhibition of SULT Activity by OH-PCBs

To assess inhibition of 3-OH-BaP SULT activity, we prepared stock solutions of OH-PCBs in dimethyl sulfoxide (DMSO) and added aliquots to incubation mixtures such that the final concentration of OH-PCB was in the range of 0.01-200 [micro]M and the DMSO concentration did not exceed 0.5% (vol/vol). For each OH-PCB, we examined the concentration dependence of inhibition with three liver cytosol fractions, as well as with cytosol fractions from the E. coli expressing SULT1A1*1, SULT1A3, SULT1B1, and SULT1E1, and the purchased Sf-9 cytosol fraction containing SULT1A1*2. For studies with HL cytosol, SULT1A1*1, and SULT1E1, we examined two concentrations of 3-OH-BaP, 0.1 [micro]M and 1.0 [micro]M.

For studies with SULT1A1*2, we examined only 0.1 [micro]M 3-OH-BaP, a concentration that did not elicit substrate inhibition. For studies with SULT1B1, we used only 1.0 [micro]M 3-OH-BaP because this enzyme had very low activity at 0.1 [micro]M 3-OH-BaP and did not exhibit substrate inhibition. Examination of the effect of 50 [micro]M concentrations of several OH-PCBs on the activity of SULT1A3, measured with 1.0 [micro]M 3-OH-BaP, revealed little inhibition, so no further concentrations were studied.

Kinetics of Inhibition

To study the type of inhibition produced by OH-PCBs, we used 4'-OH-CB112 (B5) as a model inhibitor. We prepared four sets of assay tubes containing HL cytosol and varying amounts of 3-OHBaP from 35 to 150 nM: one set (control) contained no 4'-OH-CB112; the other sets contained 0.25 [micro]M, 0.5 [micro]M, or 1.0 [micro]M 4'-OH-CB112. We calculated the enzyme kinetic parameters from studies with variable concentrations of 3-OH-BaP using nonlinear regression analysis and GraphPad 4.0 software. We selected the built-in Michaelis-Menten equation for most analyses. Where we found evidence of 3-OH-BaP substrate inhibition, we fit the data into an equation derived from a two-substrate model:

$$V = [V_1][1 + ([V_2]S/[V_1][K_i])] /(1 + [K_m]/S + S/[K_i]).$$

This equation denoted the constant for binding of the first substrate (S) molecule as Km and the second substrate molecule as $[K_i]$. $[V_1]$ is the maximum rate for the noninhibitory substrate concentration range, and $[V_2]$ is the minimum rate in the inhibitory substrate concentration range.

We calculated the effects of OH-PCBs on 3-OH-BaP SULT activity as percentage inhibition compared with the controls without an inhibitor. We obtained $I[C_{50}]$ values by fitting log OH-PCB concentration and percent control activity to a sigmoidal curve. We examined the relationships between $I[C_{50}]$ and physicochemical properties of the OH-PCBs by linear correlation analysis. We calculated the inhibitory constant ($[K_i]$) from the kinetic studies with 4'-OH-CB112 by means of Dixon plots and plots of $[K_m]/[V_{max}]$ against inhibitor concentration.

SULT1A1 Genotype of the Liver Donors

We found that the HL cytosols used were from individuals with different SULT1A1 genotypes, as determined by PCR amplification of the region of the SULT1A gene flanking the polymorphic base pair. The G to A mutation in SULT1A1 removed the restriction site for the endonuclease HaeII. An individual homozygous for the SULT1A1*2 allele did not have the HaeII restriction site, and the PCR product was not cleaved (lane 1). The PCR product from the individual homozygous for SULT1A1*1 showed complete cleavage by HaeII, generating two fragments of approximately 100 and 181 bp (lane 3). Enzymatic digestion of the PCR product from the heterozygote (SULT1A1*1/*2) generated one band of 281 bp and the two fragments of 100 and 181 bp (lane 2). Thus, the individual liver designated HL 1 was homozygous for the SULT1A1*1 allele, HL 2 was heterozygous, and HL 3 was homozygous for the SULT1A1*2 allele.

Sulphonation of 3-OH-BaP by HL Cytosol and Expressed Human SULT Isoforms

Initial studies of the sulphonation of 3-OH-BaP by HL cytosol revealed that concentrations of 3-OH-BaP > 0.15 [micro]M resulted in a decrease in activity. To find a saturating concentration of PAPS, we conducted incubations in the presence of 0.1 [micro]M 3-OH-BaP and varying concentrations of PAPS. The data fit the Michaelis-Menten equation, with an apparent [K_m] of 0.56 [+ or -] 0.09 [micro]M and a V[max of 48 [+ or -] 2 pmol/min/mg protein (mean [+ or -] SD; n = 3). The dependence of activity upon PAPS concentration in expressed human SULT1A1*2, in the presence of 0.1 [micro]M 3-OH-BaP, also followed Michaelis-Menten kinetics. The apparent [K_m] was 0.32 [micro]M, and [V_{max}] was 684 pmol/min/mg protein. Cytosol and the expressed enzyme were saturated by a PAPS concentration of 10 [micro]M, and we used this concentration in subsequent studies.

We conducted detailed studies of the effect of a range of 3-OH-BaP concentrations up to 2 [micro]M on reaction rates with HL cytosol and expressed human SULT1A1*2. We obtained preliminary estimates of the kinetic constants [K_m] and [V_1] by fitting the initial rates of sulphonation at concentrations < 0.15 [micro]M 3-OH-BaP to the Michaelis-Menten equation. We then obtained the values of [K_i] and [V_2] through constraining [K_m] using the equation of Zhang et al. (1998). We also analysed data by constraining [V_1], but a better fit was found when constraining [K_m]. Kinetic studies with expressed

SULT1A1*2 revealed substrate inhibition with the single enzyme. The expressed enzyme showed a lower value for [K_m] (0.022 [micro]M) and [K_i] (0.16 [micro]M) than did any of the HL cytosols.

The values are from substrate concentration ranges in which the data fit the Michaelis-Menten equation. SULT1A1*1 and SULT1E1 showed substrate inhibition at concentrations of 3-OH-BaP > 0.15 [micro]M, but detailed kinetic analyses at inhibitory concentrations was not conducted with these expressed enzymes. We found that SULT1A1*1 had an apparent [K_m] (0.018 [micro]M) similar to that found with SULT1A1*2 (0.022 [micro]M). SULT1E1 also had high affinity for 3-OH-BaP, with an apparent [K_m] of 0.05 [micro]M. SULT1A3 and SULT1B1 did not exhibit substrate inhibition over a concentration range up to 5 [micro]M and showed much higher apparent [K_m] values for 3-OH-BaP. These expressed enzyme preparations were partially purified, and SDS-PAGE showed they contained different percentages of the respective SULT enzymes. The values shown for [V_{max}] were corrected for the percentage of each respective SULT isoform in the partially purified enzyme preparation.

Inhibition of 3-OH-BaP Sulphonation by OH-PCBs with HL Cytosol

The 4-OH-PCBs with one (B group) or two (C group) flanking chlorine substituents inhibited HL cytosolic 3-OH-BaP sulfotransferase activity in a concentration-dependent manner. Compounds B1-B5 with the 3-chloro-4-hydroxy substitution pattern were potent inhibitors, with I[C_{50}] values ranging from 0.33 to 1.08 [micro]M, when activity was measured with 0.1 [micro]M 3-OH-BaP.

The OH-PCBs with two chlorine atoms flanking the hydroxy group (C1-C6) were less potent inhibitors under these conditions. The single 6-OH-PCB studied, A1, was a very weak inhibitor, with an I[C_{50}] of > 100 [micro]M. When activity was measured with 1 [micro]M 3-OH-BaP, a concentration at which substrate inhibition occurred, the measured I[C_{50}] values showed lower inhibitory potencies for all OH-PCBs, but especially so for the C group compounds, whose I[C_{50}] values ranged from 3 to 58.7 [micro]M.

By chance, the three HL cytosol fractions we used in these studies were from individuals with different SULT1A1 genotypes. One was SULT1A1*1 homozygous, a second was heterozygous for SULT1A1*1/*2, and the third was SULT1A1*2 homozygous. Kinetic analysis showed little difference among the three cytosol fractions for V1, which was 121 pmol/min/mg for the homozygous SULT1A1*1 liver and 94 pmol/

min/mg protein for the SULT1A1*2 liver; however, the small size of our sample precludes a more detailed analysis of genotype effects on 3-OH-BaP sulphonation activities.

In previous studies, we showed that OH-PCBs inhibited 3-OH-BaP sulphonation in catfish intestinal cytosol and that a compound structurally related to OH-PCBs, 2,4,4'-trichloro-2'hydroxydiphenyl ether (triclosan), inhibited sulphonation and glucuronidation of 3-OHBaP and other substrates in HL cytosol and with SULT1A1, SULT1B1, and SULT1E1. Here we demonstrated that a set of 4-OH-PCBs inhibited SULT activity with 3-OH-BaP, the major metabolite of another pollutant chemical, BaP, in HL cytosol as well as with cDNA-expressed SULTs.

In HL cytosol, all the 4-OH-PCBs examined inhibited the sulphonation of 3-OH-BaP. Under incubation conditions in which the 3-OH-BaP substrate did not cause substrate inhibition (0.1 [micro]M 3-OH-BaP), compounds with one chlorine atom adjacent to the OH group (B1-B5) were more potent inhibitors of sulphonation than were compounds in type C, with chlorine atoms flanking the OH group on each side. We observed very similar results for potency of inhibition and order of inhibitory potency with all three liver cytosol fractions and the two allelic variants of expressed SULT1A1.

When incubated with 1.0 [micro]M 3-OH-BaP, a concentration that produced substrate inhibition in liver cytosol and with both SULT1A1 variants, the OH-PCBs were considerably less potent inhibitors in cytosol and even more so with the expressed SULT1A1*1 and SULTIA12 enzymes. The effect of substrate concentration on the inhibitory potency of the OH-PCBs suggested the possibility that the OH-PCBs competed with the 3-OH-BaP for an inhibitory site of the SULT1A1 protein. Gamage et al. (2003) reported that SULT1A1*2 could accommodate two molecules of the xenobiotic model substrate p-nitrophenol in the active site.

They proposed that substrate inhibition at high concentrations of p-nitrophenol was due to impeded catalysis when both binding sites were occupied. The active site of SULT1A1 appears to be plastic enough to accept a wide range of hydrophobic phenolic compounds and may be able to accommodate two molecules of 3-OH-BaP, leading to substrate inhibition, or one molecule of 3-OHBaP and one molecule of OH-PCB, resulting in the OH-PCB inhibiting 3-OH-BaP sulphonation. The kinetic studies with 4'-OH-CB112 (B5) showed that the mechanism of inhibition was noncompet-itive. This result could fit the scenario for inhibition discussed above but does not suggest direct competition of the OH-PCB for binding to the active site in an orientation that favours sulphonation.

Whatever the mechanism of inhibition, the loss in inhibitory potency of OH-PCB when assays were conducted with 1.0 [micro]M 3-OH-BaP suggested that the enzyme favoured binding of 3-OH-BaP over binding of OH-PCB, and this was especially true for type C OH-PCBs, which showed a greater loss in potency than did the type B compounds. These findings suggest that OH-PCBs are likely to be poor substrates for sulphonation, but this has not been studied in human liver. We could not discern any other clear relationship of inhibitory potency with structural features or with physicochemical properties of the OH-PCBs in this relatively small series of compounds, with cytosol or the two expressed SULT1A1 enzymes. The small size of the series of compounds studied and the lack of ready availability of a systematic series of 4-OH-PCBs prevent further analysis of structure-potency relationships at this time.

Table: *Kinetic parameters for 3-OH-BaP sulphonation by human >livers and SULT1A1*2.*

[K_m]	**([micro]M)**	**[K_I]**	**([micro]M)**
>HL 1	cytosol	0.048	0.915
>HL 2	cytosol	0.051	0.534
>HL 3	cytosol	0.048	0.460
>Mean	[+ or -] SD 0.049	[+ or -] 0.01 0.636	[+ or -] 0.244
>SULT1A1*2		0.022	0.160
	[V_1] (pmol/min/mg protein)	**[V_2] (pmol/min/mg protein)**	**[R^2]**
>HL 1 cytosol	121	2	0.95
>HL 2 cytosol	93.0	38.50	0.956
>HL 3 cytosol	94.4	31.0	0.958
>Mean	[+ or -]SD	102 [+ or -]	15.8 28.3 [+ or -]19.3
>SULT1A1*2	4,400	290	0.953

Kinetic analysis was performed using a two-substrate model as described in "Materials and Methods." HL 1 cytosol was homozygous for SULTIA1*1, HL 2 cytosol was heterozygous as SULT1A1*1/*2, and HL 3 cytosol was homozygous for SULT1A1*2.

Of the other expressed enzymes studied, only SULT1E1 exhibited potent inhibition by the 4-hydroxylated PCBs. The structure-inhibitory potency requirement for SULT1E1 was very different from that for HL cytosol, SULT1A1*1, or 1A1*2, where type B compounds were more potent inhibitors than were type C OH-PCBs. With SULT1E1, OH-PCBs with no or one ortho-substituted chlorine were more potent as inhibitors of 3-OH-BaP sulphonation than were those with two ortho-substituted chlorine atoms. Substituted biphenyls with less than one ortho substituent preferentially adopt coplanar conformation of the two phenyl rings, whereas those with two or more ortho substituent atoms take on noncoplanar conformations. We found a significant linear correlation between inhibitory potency and calculated solution dihedral angles. Similarly, Kester et al. (2000) found that the best OH-PCB inhibitors of estrogen sulphonation (I[C_{50}] values < 5 nM) did not have chlorine substituents at the 2- or the 6-position. Shevtsov et al. (2003) later showed that 4,4'-di-OH-CB80 (4,4'-diOH-3,3',5,5'-tetrachlorobiphenyl) did not bind the SULT1E1 in a planar conformation, but rather with a 30° twist between the phenyl rings. We found that the four OH-PCBs with solution dihedral angles of 38° were more potent inhibitors than were those with larger dihedral angles. Although it is possible that interaction with the protein could alter the conformation of the OH-PCBs, resulting in a different dihedral angle for the enzyme-bound OH-PCB, our results show that lack of ortho substituents is associated with higher inhibitory potency for a xenobiotic SULTIE1 substrate, 3-OH-BaP.

Table: *Apparent kinetic constants for cDNA-expressed >sulfotransferase with 3-OH-BaP as substrate.*

SULT	*[V.sub.max] [K.sub.m] ([micro]M)*	*[V.sub.max]/[K.sub.m] (nmol/min/ mg protein)*	*(mL/min/ mg protein)*
SULT1A1*1	0.018	6.89	383
SULTIA3	2.90	333.3	115
SULT1B1	2.00	9.70	4.9
SULT1E1	0.05	8.35	167

Partially purified SULT isoforms were used for these studies.[V_{max}] was calculated from the mg/mL of the partially purified preparation and corrected by the percentage of protein estimated to be SULT, from SDS-PAGE: 44.0% for SULT1A1*1, 39.0% for SULT1A3, 80.3% for SULT1B1, and 65.0% for SULT1E1.

SULT1A3 metabolised 3-OH-BaP with a very high [V_{max}], although its preferred substrates are reported to he catecholamines and other monocyclic phenols containing hydrogen bond donors. Interestingly, 50 [micro]M OH-PCBs caused little or no inhibition of this enzyme, thereby showing that the inhibitory interaction was enzyme selective. SULT1B1, the thyroid hormone sulfotransferase, catalysed the sulphonation of 3-OH-BaP; however, OH-PCBs that were potent inhibitors of SULT1A1 were only weak inhibitors of the SULT1Bl-catalysed reaction. In contrast to results with the other enzymes, compound A1 (6'-OH-CB35) was a fairly potent inhibitor of SULT1B1. Previously, ortho-, meta-, and para-hydroxylated PCBs were found to inhibit thyroid hormone sulphonation. The meta-hydroxylated PCB, 3-OH-2,3',4,4',5-pentachlorobiphenyl (3-OH-CB118), was the most potent inhibitor of thyroid hormone sulphonation in male rat liver cytosol, followed by two para-hydroxylated PCBs. The ortho-hydroxylated PCB had the lowest potency among the four OH-PCBs studied. However, with 3-OH-BaP as substrate, the ortho-OH-PCB, 6'-OH-CB35, was a more potent inhibitor than were those with para-OH groups, which suggested that the inhibitory interaction with SULT1B1 was substrate dependent.

Table: *In vitro inhibition of 3-OH-BaP sulfotransferase activity by the tested OH-PCBs using HL cytosol and cDNA-expressed sulfotransferases at 0.1 and 1.0 [micro]M substrate concentratinn. (a)*

Compound	***Dihedral Compound***	***Log D at pH 7.0***	***Angle ([degrees])***
no.			
A1	6'-OH-CB35	4.7	50
B1	4'-OH-CB35	4.7	38
B2	4'-OH-CB36	4.8	38
B3	4'-OH-CB69	5.1	72
B4	4'-OH-CB106	5.2	60
B5	4'-OH-CB112	5.2	78
C1	4'-OH-CB79	4.5	38
C2	4-OH-CB36	4.2	38
C3	4'-OH-CB121	4.7	72
C4	4'-OH-CB159	4.7	78
C5	4'-OH-CB165	4.6	78
C6	4'-OH-CB72	4.5	57
	I[C.sub.50] ([micro]M) 0.1 [micro]M 3-OH-BaP		

Compound no.	*HL cytosol*	*SULT1A1*1*	*SULT1A1*2*	*SULTIE 1*
A1	> 100	> 100	> 100	~100
B1	0.33 [or -] 0.02	0.77	0.55	0.24
B2	0.67 [or -] 0.12	1.31	0.94	0.45
B3	0.91 [or -] 0.09	1.16	1.31	4.87
B4	0.37 [or -] 0.04	1.07	1.06	1.18
B5	1.08 [or -] 0.12	1.17	1.48	5.35
C1	6.71 [or -] 0.91	6.65	4.57	0.50
C2	2.30 [or -] 0.45	3.09	3.05	0.41
C3	3.95 [or -] 0.23	8.15	6.52	7.98
C4	1.31 [or -] 0.14	2.15	1.67	1.30
C5	2.87 [or -] 0.09	2.58	2.59	6.96
C6	1.72 [or -] 0.21	2.21	2.03	0.57

I[C.sub.50] ([micro]M)

1.0 [micro]M 3-OH-BaP

Compound no.	*HL cytosol*	*SULT1A1*1*	*SULT1B1*	*SULTIE1*
A1	> 100	> 100	4.72	> 100
B1	0.96 [or -] 0.30	25.2	16.8	1.02
B2	1.05 [or -] 0.39	28.0	37.0	1.89
B3	1.50 [or -] 0.32	67.5	> 100	30.8
B4	2.51 [or -] 0.67	10.3	17.4	6.97
B5	4.22 [or -] 1.03	42.5	86.5	23.2
C1	58.7 [or -] 13.9	59.8	39.9	1.32
C2	35.9 [or -] 1.47	> 100	47.5	1.65
C3	44.6 [or -] 6.42	99.5	> 100	16.7
C4	38.4 [or -] 15.2	34.1	> 100	3.55
C5	47.3 [or -] 10.2	54.8	> 100	21.4
C6	3.05 [or -] 0.41	33.8	> 100	2.28

(a) Values for HL cytosol are the means [or -] SDs of three livers, tested in duplicate; results for expressed SULT enzymes are the means of duplicate determinations.

Table *: Apparent kinetic constants for 3-OH-BaP sulfotransferase activity in HL cytosol in the presence and absence of 4'-OH-CB112 (B2).*

B5 ([micro]M)	*[K.sub.m] ([micro]M)*	*[V.sub.max](pmol/min/ mg protein)*
0	0.045 [or -] 0.02 A	69.1 [or -] 8.0 B
0.25	0.043 [or -] 0.02 A	56.3 [or -] 6.0 B
0.5	0.045 [or -] 0.01 A	41.2 [or -] 3.3 C
1.0	0.066 [or -] 0.02 A	33.3 [or -] 3.2 D

* Values for liver cytosol are mean = SD (n = 3), except for studies with 0.25 [micro]M B5, where two livers were used. Different letters indicate values that are significantly different from each other (P < 0.05).

levels. Type B OH-PCBs with the 3-chloro4-hydroxy substitution pattern do not appear to be persistent in blood; however, of the 209 PCB congeners, 19 have a 3-chloro substitution in one of the phenyl tings, which can be biotransformed to type B OH-PCBs. If type B OH-PCBs are formed in people, their high potency as inhibitors of 3-OH-BaP sulphonation may cause increased toxicity in people who are coexposed to PAH and PCBs.

We found that several OH-PCBs, especially those with a 3-chloro-4-hydroxy substitution pattern in the phenolic ring, inhibited the sulphonation of 3-OH-BaP in cytosol and with SULT1A1 at submicromolar concentrations. Some OH-PCBs with no or one ortho chlorine were potent inhibitors of 3-OH-BaP sulphonation with SULT1E1. SULT1B1- and SULT1A3-catalysed sulphonation of 3-OH-BaP was less sensitive to inhibition by OH-PCBs. The inhibitory interaction of OH-PCBs with SULT1AI and SULT1E1 may have consequences for the biotransformation and toxicity of phenolic xenobiotics.

Neuroinflammation

Idiopathic Parkinson's disease (PD) is a devastating movement disorder characterised by selective degeneration of the nigrostriatal dopaminergic pathway. Neurodegeneration usually starts in the fifth decade of life and progresses over 5-10 years before reaching the fully symptomatic disease state. Despite decades of intense research, the etiology of sporadic PD and the mechanism underlying the selective neuronal loss remain unknown. However, the late onset and slow-progressing nature of the disease has prompted the consideration of environmental exposure to agrochemicals, including pesticides, as a risk factor. Moreover, increasing evidence suggests that early-life occurrence of inflammation in the brain, as a consequence of either brain injury or exposure to infectious agents, may play a role in the pathogenesis of PD. Most important, there may be a self-propelling cycle of inflammatory process involving brain immune cells (microglia and astrocytes) that drives the slow yet progressive neurodegenerative process. Deciphering the molecular and cellular mechanisms governing those intricate interactions would significantly advance our understanding of the etiology and pathogenesis of PD and aid the development of therapeutic strategies for the treatment of the disease.

Parkinson's disease (PD) is a degenerative neurologic disorder characterised by progressive degeneration of the nigrostriatal dopaminergic pathway that regulates body movements. Degeneration of the nerve terminals in the striatum (ST) and the neuronal cell bodies of dopamine-containing neurons in the substantia nigra (SN) eventually leads to the development of movement disorders, including resting tremor, rigidity, bradykinesia, and gait disturbance. In addition to the degeneration of dopaminergic neurons in the SN, postmortem analysis of the brains of PD patients has frequently detected the existence of cytoplasmic inclusions in the SN neurons, also known as Lewy bodies. Formation of Lewy bodies is a pathologic hallmark of PD and an affirmative postmortem diagnostic marker.

Ever since its initial description by British physician James Parkinson in 1817, the etiology of PD as well as the precise mechanism of action underlying the selective destruction of the nigrostriatal dopaminergic pathway has remained unknown. Currently, clinical diagnosis of PD is based on movement-related behavioural abnormalities. However, the absence of distinct biomarkers for the prognosis of the disease, because of the lack of clear understanding of the disease process, has significantly hampered efforts to identify and treat PD patients early in the disease course. At the present time, dopamine replacement therapy using levodopa is the most widely used approach in the treatment of clinically diagnosed and often advanced PD patients. However, the efficacy of such neurotransmitter replacement therapy has been limited because of resistance and loss of response.

Advances made in the last several decades at the forefront of molecular cloning, biochemical characterisation, pathology analysis, and epidemiologic studies have come to the general consensus that most (> 95%) PD cases are sporadic and have a late onset. A small faction (< 5%) of cases is characterised by early onset, and these mostly occur in familial clusters. Development of parkinsonism syndrome in those individuals has been attributed to mutations in several recently identified genes, including parkin and [α]-synuclein. In contrast, development of idiopathic PD may represent the final outcome of a complex set of interactions among the innate vulnerabilities of the nigrostriatal dopaminergic system, potential genetic predisposition, and exposure to environmental toxicants. There is growing recognition of the role of inflammation in the brain (neuroinflammation) in the pathogenesis of PD. Neuroinflammation in the brain can be induced by exposure to either infectious agents or toxicants. Neuroinflammation can also occur as a sequela to neurotoxicant-elicited neuronal damage or injuries in the brain, a process called reactive gliosis. In this chapter, we review the literature on the potential impact on the development of idiopathic PD of environmental factors such as exposure to infectious agents and pesticides and early-life occurrence of brain injuries. More important, we will attempt to establish a possible link between early-life neuroinflammation and late life development of idiopathic PD.

Immune Cells in the Brain

Inflammation in the brain involves primarily the activity of two types of glial cells: microglia and astrocytes. Microglia are the resident immune cells in the brain. During late embryonic and early postnatal brain remodelling and maturation, microglia are involved in the programmed elimination of neural cells.

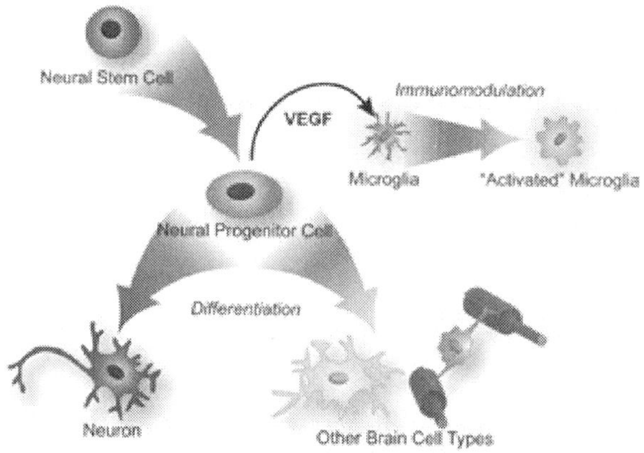

In mature brains, resting microglia exhibit a characteristic ramified morphology and serve the critical role of immune surveillance. As an important line of defence in the brain, microglia become readily activated in response to injuries to the brain or to immunologic stimuli. Activated microglia undergo dramatic morphologic changes, metamorphosing from resting ramified microglia into activated amoeboid microglia. They also exhibit increased expression of surface molecules such as complement receptors and major histocompatibility complex (MHC) molecules. At the same time, activated microglia release a variety of soluble factors. Although activated microglia are known to produce several trophic factors, most of the factors released by activated microglia are proinflammatory and potentially cytotoxic.

Astrocytes, under physiologic conditions, provide glia-neuron contact, maintain ionic homeostasis, buffer excess neurotransmitters, and secrete neurotrophic factors. In response to immunologic challenges and brain injuries, astrocytes also become activated. In the course of activation, astrocytes up-regulate the expression of cell-type-specific proteins such as the glial fibrillary acidic protein. In addition, they secrete neurotrophic factors and several proinflammatory cytokines.

Microglia as the Primary Contributor of Proinflammatory and Neurotoxic Factors

In terms of the release of proinflammatory and cytotoxic factors, activated microglia differ significantly from activated astrocytes in several respects.

First, the repertoire of factors produced by activated microglia is far more complex than that of activated astrocytes. For example, bacterial endotoxin lipopolysaccharide (LPS)-stimulated production of tumour necrosis factor-[α] (TNF-[α]) appears to be limited to microglia and has not been observed in astrocytes. Astrocytes, but not microglia, failed to respond to [β]-amyloid peptide-stimulated production of superoxide free radicals. Second, of the factors produced by both cell types, microglia produces a significantly larger quantity of most such factors than do astrocytes. For instance, LPS-stimulated microglia produced several-fold higher amounts of nitrite, an indicator of the production of nitric oxide (NO), than did LPS-stimulated astrocytes. Similarly, the LPS-stimulated production of interleukin-1[β] (IL-1[β]) was far more prominent in microglia than in astrocytes.

A notable exception may be that activated astrocytes, as well as activated microglia, produce abundant quantities of prostaglandins (PGs), such as PGE and PGD. Third, in response to immunologic stimuli, the production kinetics of proinflammatory factors is different. Microglia usually responds faster than astrocytes.

For example, LPS-induced production of NO occurred within the first 6-12 hr in microglia, compared with a much delayed time point (24 hr) in astrocytes. Interestingly, astrocytes appear to rely on secondary responses to produce certain proinflammatory factors. For example, human astrocytes that were insensitive to LPS stimulation responded well to stimulation by IL-1[β] and produced fair quantities of TNF-[α] and IL-6. These observations strongly imply that microglia, as the first line of response to immunologic challenges in the brain, play the primary role in the brain inflammatory process and therefore are, under pathologic conditions, the predominant contributor to the inflammation-mediated neurodegenerative process.

Overproduction and Accumulation of Proinflammatory and Neurotoxic Factors Are Deleterious to Neurons

Numerous in vitro and in vivo studies have demonstrated that the production and accumulation of proinflammatory and cytotoxic factors by activated glia have an impact on neurons, inducing neurodegeneration. Neurotoxicity has been attributed to high levels of NO, IL-1[β], IL-6, TNF-[α], and reactive oxygen species (ROS) such as superoxide anions. Besides being individually toxic to neurons, NO and superoxide may form more toxic intermediates such as peroxynitrite. Individual factors also may work in concert to induce synergistic neurotoxicity, a scenario that may bear a closer resemblance to the in vivo neurodegenerative process. Conflicting results have been reported regarding the involvement of prostanoids in inflammation-mediated neurodegeneration. On the one hand, inhibitors of cyclooxygenases and/or 5-lipoxygenase have been shown to protect neurons from inflammation-mediated toxicity. On the other hand, exogenous PG[E_2] has been reported to attenuate LPS-induced neurotoxicity. Nevertheless, it is generally believed that the accumulation of a variety of proinflammatory and neurotoxic factors released from activated microglia eventually results in neurodegeneration.

In reference to dopaminergic neurodegeneration, it is especially important to note that the SN dopaminergic neurons are characteristically sensitive to insults by a variety of external factors because of their reduced antioxidant capacity; their high content of dopamine, melanin, and lipids that are prone to oxidation; and potential defects in mitochondrial function. Furthermore, the SN area of the brain is particularly rich in microglia. Hence, SN dopaminergic neurons, residing in a microglia-rich environment, are especially vulnerable to attacks imposed by factors produced by activated microglia.

Findings Associating Brain Inflammation with the Pathogenesis of PD

The initial findings linking inflammation in the brain with the pathogenesis of PD have been derived from postmortem analysis of the nigra of PD patients. In 1988, McGeer et al. reported the detection of large populations of MHC [human leukocyte antigen (HLA)-DR]-positive reactive microglia in the SN of the PD brains they analysed. Since then, numerous studies have reported the detection of significantly elevated levels of a variety of proinflammatory factors in the SN, ST, or cerebrospinal fluid. Those factors included components of the innate immune response such as complement proteins and cytokines such as IL-1, IL-2, IL-6, and TNF-[α]. In addition, inducible NO synthase (iNOS)-positive glial cells also have been detected in the midbrain region of PD brains. These observations have clearly demonstrated that inflammation, particularly microglial activation, is involved in the pathogenesis of PD. In addition, molecular genetic analysis of polymorphisms in the genes for IL-1, IL-6, and TNF-[α] as well as the TNF-[α] receptor gene of PD patients and matched controls further supports the involvement of cytokines in the pathogenesis of PD. Nevertheless, because most of the changes were detected in the terminal stage of the disease, it has remained an open question whether the observed glial activation was merely a consequence of the reactive glial response to neuronal loss. Clearly, additional evidence will be needed to establish that glial activation, especially microglial activation, is involved in an earlier stage of dopaminergic neurodegeneration during the PD pathogenetic process.

Is Exposure to Infectious Microorganisms a Risk Factor for PD? For several decades, infectious agents have been suspected to be risk factors for PD. Epidemiologic studies have associated early-life viral infections to postencephalitic PD; case reports, on the other hand, tend to relate viral infections to the development of acute parkinsonism. Experimentally, exposure of neonatal Fisher 344 rats to Japanese encephalitis virus can induce the degeneration of SN dopaminergic neurons, development of movement disorders resembling human PD, and occurrence of gliosis in the SN. However, identification of a specific causative viral agent (DNA or antigens) in humans has so far proven elusive. It is possible that viral infections alone are not sufficient to initiate the development of idiopathic human PD, but viral infection-associated inflammation at an early stage of life may well play a role in the pathogenesis of the disease. However, the probability of unearthing "footprints" (i.e., viral DNA and/or antigens) of early-life viral infections in postmortem analyses of PD brains might be exceptionally low.

Besides viruses, infections by bacterial pathogens have also been proposed to play a role in the development of PD. Because of the increased frequency of peptic ulcers observed in some PD patients, infection by Helicobacter pylori has been one such candidate bacterium. Small-scale paired-case studies appear to support the hypothesis. However, solid evidence and experimental animal models are lacking to establish a role for H. pylori infection in the pathogenesis of PD. Of particular interest, however, is the demonstration by Kohbata and Beaman (1991) that mice injected with Nocardia asteroides, a common bacterium in the soil, developed PD-like syndromes. Although the relevant mechanism of action remains unknown, subsequent in vitro and in vivo studies showed that infection with one strain of Nocardia (GUH-2) resulted in apoptotic death of SN dopaminergic neurons. Although the involvement of glial cells in the neurotoxic process was not established and a serologic case-control study could not associate Nocardia exposure to human PD, continued research is warranted.

Inflammogen-Induced Dopaminergic Neurodegeneration: Experimental Models of Inflammation-Related PD

To directly test the hypothesis that inflammation in the brain can result in selective dopaminergic neurodegeneration, a number of groups have used LPS to evoke inflammation in the brain. Initially, LPS was injected directly in a bolus manner into the SN area of rat brains. Dramatic degeneration of the nigrostriatal dopaminergic pathway was observed in those animals receiving the LPS injections. However, this single application-induced neurodegeneration was not selective to dopaminergic neurons. More important, inflammation (e.g., microglial activation) induced by the relatively large quantity of LPS (microgram levels) was too fast and too robust to enable us to clarify the temporal relationship between microglial activation and dopaminergic neurodegeneration. A chronic LPS infusion model was subsequently established. In this model, LPS was chronically infused for 2 weeks at a rate of 5 ng/hr into the SN region of rat brains using an epidermal osmotic minipump.

Maximal microglial activation occurred in the first 2 weeks, but significant degeneration of SN dopaminergic neurons did not manifest until 4-6 weeks after the LPS infusion. In in vitro studies using mesencephalic neuron-glia cultures, inhibition of microglial activation and reduction of the production of microglia-derived neurotoxic factors significantly attenuated the LPS-induced dopaminergic neurodegeneration. In addition to infusion of LPS into adult rodent brains, in utero exposure of developing fetuses to LPS results in degeneration of the nigrostriatal dopaminergic pathway in neonates. These studies thus provide a tentative mechanistic link between the occurrence of inflammation in the brain and dopaminergic neurodegeneration and also provide experimental evidence to suggest that the occurrence of neuroinflammation in early life, if not an inducer by itself, may represent a risk factor for development of PD later in life.

Traumatic Brain Injury and Development of PD

As with the potential association of infection with development of PD, the possible correlation between the occurrence of closed head injury with PD development rests largely on information obtained through epidemiologic studies and physicians' case reports. Analysis of the medical histories of a large group of World War II veterans pointed to an association between the occurrence of severe antecedent head injuries and the development of PD in later life. This notion is supported by several other studies, including several case-control studies, that have identified head injury as a major risk factor for the development of PD. In contrast, several other studies failed to establish a correlation between the occurrence of head injuries and later development of PD. This contradiction may reflect the difficulty of designing large-scale epidemiologic studies, identifying reliable parameters for measurements, and interpreting complex data. Nevertheless, numerous case reports have described episodes of traumatic brain injuries with the occurrence of an acute-phase parkinsonism or later life development of PD. Worth noting is the hypothesis of so-called "Boxer's parkinsonian syndrome," which proposes that boxers, because of their professionally related higher chance of suffering head injuries, appear to have a higher incidence of Parkinson-like movement disorders.

Although the mechanism underlying this head trauma-related development of permanent damage to the nigrostriatal pathway remains unclear, it is possible that the initial neuronal damage leads to the activation of glial cells, whose activity may exacerbate the neurodegenerative process. This may be especially true in late-onset cases of PD after brain injuries. Using an animal model of closed head injury, Shohami et al. (1997) have demonstrated significant elevations in the levels of a number of cytokines (including IL-1, IL-6, and TNF-[α]) after head injury in rats. The production and accumulation of those proinflammatory factors may exacerbate the neuronal damage initially induced by the head injury.

A fundamental yet unsolved issue, however, is the identity of the responsible initiating factor(s) that is presumably released from injured neurons and that stimulates the activation of glial cells. Over the years, a variety of soluble factors as well as cell-cell adhesion molecules has been proposed and/or implicated. Continued search for such factors and elucidation of the molecular mechanism responsible for the neuronal injury-induced reactive gliosis will advance our understanding of the complex pathogenesis of PD.

Exposure to Pesticides and Development of PD

Exposure to agrochemicals, particularly pesticides, has long been suggested as a risk factor for PD. The identification of 1-methyl-4-phenyl-1,2,3-tetrahydropyridine (MPTP), a by-product of illicit heroin synthesis, as the culprit that induced Parkinson syndromes in humans has significantly intensified the search for environmental factors as potential causes of PD. To date, exposure to several classes of pesticides has been reported to result in dopaminergic neurotoxicity in animal models, and such pesticides have hence been proposed as potential risk factors in humans.

The structural similarity between 1-methyl-4-phenylpyridinium ion (MP[P^+]), the active metabolite of MPTP, and a common herbicide, 1,1'-dimethyl-4,4'-bypyridinium (paraquat), prompted speculation that paraquat might be a dopaminergic neurotoxicant. Barbeau et al. (1985) reported that paraquat, like MPTP, induced parkinsonian behavioural changes and dopamine depletion in the northern leopard frog (Rana pipiens). Subsequently, it was reported that exposure to pesticides including paraquat was positively correlated with increased incidence of PD in patients in parts of Canada. Similar observations were reported in farming communities in Taiwan. Research conducted in the last decade has provided a more in-depth understanding of the dopaminergic neurotoxicity of paraquat. Intracerebral injection of paraquat resulted in loss of SN dopaminergic neurons, depletion of dopamine in the SN, and an elevated response to apomorphine-induced rotational bahaviour. The reported effects of systemically administered paraquat, on the other hand, have been less consistent. For example, Brooks et al. (1999) reported that systemic administration of paraquat to C57BL/6 mice resulted in the loss of SN dopaminergic neurons, degeneration of ST dopaminergic fibres, and reduced ambulatory activity. In contrast, McCormack et al. (2002) reported that mice that underwent repeated systemic administration of paraquat exhibited a selective loss of SN dopaminergic neurons without a reduction in ST dopamine content. Instead, enhanced dopamine synthesis in the ST, which was attributed to a possible compensatory mechanism, was observed. Mechanistically, the ability of paraquat to induce free radical formation, facilitate [α]-synuclein fibrillation, and induce apoptotic cell death has been associated with its dopaminergic neurotoxicity. Another class of agrochemicals that have been found to possess dopaminergic neurotoxicity is dithiocarbamate-based fungicides. Case reports initially associated exposure to maneb, a widely used fungicide in this class, with the development of parkinsonism in humans. A rat model was recently created by direct infusion of maneb into the lateral ventricles. In vitro studies indicate that the dopaminergic neurotoxicity of maneb may be associated with its ability to inhibit the activity of complex III in the mitochondrial respiratory chain. Fitsanakis et al. (2002) had earlier proposed that the dopaminergic neurotoxicity of maneb might be associated with its ability to facilitate catecholamine oxidation. Furthermore, besides being toxic to dopaminergic neurons by itself, repeated systemic administration of maneb and paraquat to mice induced a synergistic reduction in ST dopamine content, degeneration of SN dopaminergic neurons, and development of motor behavioural abnormalities. Neonatal exposure to both maneb and paraquat also increased the susceptibility of the nigrostriatal dopaminergic system to rechallenge with the same agents during adulthood.

A third class of insecticides whose environmental exposure has been associated with increased incidence of PD includes the naturally occurring and commonly used insecticide rotenone. Rotenone is known to be a specific inhibitor of mitochondrial complex I. Because a defect in mitochondrial complex I is a well-documented feature of idiopathic PD, rotenone was suspected to be a dopaminergic neurotoxin. Intracerebral application of rotenone had earlier been shown to damage the nigrostriatal dopaminergic pathway in rats. Initial attempts to systemically administer rotenone to rats resulted in damage to ST dopaminergic fibres but not SN dopaminergic neurons. Later attempts to acutely or subchronically administer rotenone to mice also produced only minimal damage to dopaminergic neurons. In the last several years, however, several groups have demonstrated that continuous systemic administration of rotenone to rats reproduces key features of PD, including selective degeneration of the nigrostriatal dopaminergic system, formation of cytoplasmic inclusions in SN neurons, and movement disorders. In addition, chronic and subcutaneous administration of rotenone resulted in a highly selective dopaminergic damage and [α]-synuclein aggregation. The lack of agreement among these studies may stem from differences in the route, dosage, and/or duration of rotenone application. Mechanistically, although earlier studies had implied that rotenone might interfere with the dopamine transport mechanism in the ST, recent studies suggest that inhibition of mitochondrial complex I activity and facilitation of [α]-synuclein aggregation may be closely associated with rotenone's selective dopaminergic neurotoxicity. It remains to be determined whether typical human exposure to pesticides alone is likely to account for the many cases of PD. Of particular relevance may be the synergistic dopaminergic neurotoxicity observed with the combination of low levels of multiple neurotoxicants, including rotenone and LPS and paraquat and maneb.

The fourth class of pesticides with potential dopaminergic neurotoxic effect consists of the organochlorine pesticides that include dieldrin. The neurotoxic and dopamine-depleting effects of dieldrin were initially observed in intoxicated ducks, doves, and rats. Elevated levels of residual dieldrin also have been detected in the brains of PD patients. In primary mesencephalic cultures, dieldrin exhibited a selective dopaminergic neurotoxicity that might be mediated by its ability to induce ROS formation and lipid peroxidation as well as by its ability to promote [α]-synuclein fibril formation. In addition to the above-mentioned organochlorine pesticides, case reports have associated development of parkinsonism with exposure to organophosphorus insecticides. Administration of dichlorvos or chlorpyrifos to rats reduced catecholamine content and locomotor activity. It remains to be determined whether organophosphates, as well-characterised inhibitors of acetylcholinesterase, possess any selective dopaminergic neurotoxicity.

Although not used as insecticides, the impact of polychlorinated biphenyls (PCBs), widely used in industrial settings, on the dopaminergic system is worth mentioning. Postmortem analysis has revealed elevated levels of PCBs in PD brain. As with dieldrin, PCBs were found to deplete dopamine in cell cultures and in animals. The sheer number of the PCB structurally related analogs and the broad range of their systemic toxicity have made it difficult to mechanistically associate a particular variety of PCB with selective dopaminergic toxicity. Two potentially important issues concerning exposure to pesticides and the development of PD in humans should be addressed. First, elevated levels of certain pesticides found in the brains of PD patients may be merely a reflection of their rural living and well-water drinking. It does not constitute a cause-effect relationship between pesticide exposure and development of PD. Experimental models using individual pesticides have been extremely valuable for us to evaluate the neurotoxic effects of those toxicants at relatively high doses. Perhaps more relevant to the probing of the impact of pesticides on the dopaminergic system in humans are the studies in which multiple agents were employed at subtoxic concentrations, such as the combination of rotenone and LPS and of paraquat and maneb. Clearly, additional in vitro and in vivo studies and possibly epidemiologic studies in this direction will be most informative.

Table: *Detected indicators for microglial activation in the postmortem analysis PD brains.*

Marker	***Location***
HLA-DR-positive microglia	SN
iNOS-positive glia	SN
Complement-positive Lewy bodies	SN
iNOS	SN
IL-1	SN/ST/CSF
IL-2	ST/CSF
IL-6	ST/CSF
TNF-[alpha]	SN/ST/CSF
Interferon [gamma]	SN

The second issue relates to the selective dopaminergic neurodegeneration as a result of systemic exposure to pesticides. Current explanations for the elevated vulnerability of SN dopaminergic neurons, compared with neurons in other regions, to oxidative insults include a reduced antioxidant capacity, increased content of iron, high content of oxidation-prone dopamine, and potential defect in mitochondrial complex I. It remains to be determined why dopaminergic neurons in the SN, but not those in the vicinity (i.e., ventral tegmental area), are first and foremost lost in the course of the degenerative process. One possibility could that the nature and density of microglia in both regions may be different—that those in the SN are more responsive (in addition to a higher density) to stimulation by neurotoxicants or to neuronal injuries induced by the initial assault on neurons.

Therefore, the microgliosis, either as a result of direct stimulation or as a consequence of neuronal injury, presents a much more harmful microenvironment to dopaminergic neurons in the SN. By the same token, neurons in other brain regions are less sensitive to the same insults, compared with SN dopaminergic neurons. Although experimental data are lacking to demonstrate the nature of microglia in different brain regions, studies have shown that, at least in rodents, the midbrain region has a higher density of microglia than other regions examined.

The Role of Glial Cells in Pesticide-Induced Dopaminergic Neurodegeneration

Research on the dopaminergic neurotoxicity of pesticides has traditionally focused on the direct effect of those agents on neurons. However, it has recently been demonstrated that microglia actively participate in the rotenone-induced degeneration of dopaminergic neurons in mesencephalic neuron-glia cultures. In the absence of microglia, rotenone exhibited a markedly reduced toxicity to dopaminergic neurons. In the presence of microglia, however, low nanomolar concentrations of rotenone were able to activate microglia to release superoxide free radicals and facilitate the degeneration of dopaminergic neurons.

Further studies using enzyme inhibitors suggested that the rotenone-induced release of superoxide was mediated by microglial NADPH oxidase, a primary source of superoxide generation in the immune cells of both the peripheral and the central nervous systems. The detailed molecular mechanism responsible for this rotenone-stimulated activation of microglial NADPH oxidase, however, remains to be elucidated. Nevertheless, the observation of an active participation of microglia has significantly advanced our understanding of the dopaminergic neurotoxicity of rotenone, because rotenone had previously been thought to act exclusively as an inhibitor of neuronal mitochondrial complex I. Neuronal death was thought to be a consequence of the inhibition of mitochondrial complex I, which then led to a reduction in the energy supply and subsequent collapse of the mitochondrial membrane potential. In addition to the in vitro observation for the active participation of microglia in rotenone neurotoxicity, in rotenone-infused rats microglial activation has recently been reported to precede dopaminergic neurodegeneration.

Besides being able to activate microglia to release free radicals, rotenone and LPS have recently been reported to exert synergistic dopaminergic neurotoxicity. Exposure to a combination of individually nontoxic concentrations of rotenone and LPS has significantly increased dopaminergic neurodegeneration in mesencephalic neuron-glia cultures.

The observed synergistic neurotoxicity between a pesticide (i.e., rotenone) and an inflammogen (i.e., LPS), together with observations of synergistic neurotoxicity among different classes of pesticides (i.e., paraquat and maneb), is consistent with a multifactorial hypothesis of PD etiology. Additional studies are needed to examine the effect of exposure to multiple environmental neurotoxicants on the nigrostriatal dopaminergic pathway.

Microglial Activation and Reactive Microgliosis: A Self-Propelling Cycle of Neuroinflammation that Fuels the Progressive Dopaminergic Neurodegeneration

The potential role that microglia play in the dopaminergic neurodegenerative process suggests that their contribution to the death of dopaminergic neurons includes the release of a variety of neurotoxic factors in response to immunologic stimuli. Similar to macrophages, their peripheral counterpart, microglia, as the first line of defence in the brain, can be directly activated by invading pathogens.

Table: *Pesticides and dopaminergic neurotoxicity.*

Pesticide	***Features of toxicity***	***Animal model***
Pyridinium (paraquat)	DA depletion, TH-positive neuron loss, bahaviour	Rat
Dithiocarbamate (maneb)	DA depletion, TH-positive neuron loss, bahaviour	Rat
Complex inhibitor (rotenone)	DA depletion, TH-positive neuron loss, bahaviour	Rat, mouse
Organochlorine (dieldrin)	DA depletion	

Abbreviations: DA, dopamine; TH, tyrosine hydroxylase.

On the other hand, it is well documented that microglia become activated after the occurrence of neuronal injuries. Although a variety of soluble factors released from injured neurons have been proposed to be the potential stimulators of reactive microgliosis, loss of cell-cell contact between neurons and glial cells may also result in microglial activation. For example, Chang et al. demonstrated that neurons act to suppress the reactivity of glial cells through a potential dimerisation of the neural cell adhesion molecules expressed on the surface of both cell types. Neurons may be able to negatively regulate the reactivity of glial cells through other types of cell-cell contact such as the CD200 receptor-ligand interaction, analogous to CD200's role in the negative regulation of myeloid cells. Therefore, exposure to infectious agents such as viruses and bacteria can directly lead to the activation of glial cells, especially the microglia.

On the other hand, glial cells also can be activated as a reactive response to neuronal injuries inflicted by injuries or agents such as certain pesticides that can directly damage neurons. Furthermore, certain environmental toxicants, such as rotenone, can induce neuronal damage and at the same time activate microglia. In essence, regardless of the nature of the initiating factors, a cycle may exist: Microglial activation leads to neurodegeneration; neuronal injury, in turn, leads to reactive glial activation, which further exacerbates neurodegeneration. This self-propelling cycle of brain inflammation may be particularly damaging to the SN dopaminergic neurons, which reside in a microglia-rich region in the brain and which are particularly vulnerable to oxidative stress. The continuing presence of this cycle over time—plus the possible involvement of additional uncharacterised factors, intrinsic and/or over many years—leads to a massive degeneration of the nigrostriatal dopaminergic pathway and the development of symptomatic PD. Idiopathic PD is an age-related disease that may represent the final outcome of interactions among a variety of intrinsic and environmental factors. Inflammation in the brain clearly plays an active role in the pathogenesis of PD. Understanding the nature of those complex interactions may be a prerequisite for success in our quest for effective strategies to halt the progressive neurodegenerative process of idiopathic PD.

Systems Genetics

Alcoholism is a common disease resulting from the complex interaction of genetic, social, and environmental factors. Interest in the high heritability of alcoholism has resulted in many studies of how single genes, as well as an individual's entire genetic content (i.e, genome) and the proteins expressed by the genome, influence alcoholism risk. The use of large-scale methods to identify and characterise genetic material (i.e, high-throughput technologies) for data gathering and analysis recently has made it possible to investigate the complexity of the genetic architecture of susceptibility to common diseases such as alcoholism on a systems level. Systems genetics is the study of all genetic variations, their interactions with each other (i.e., epistasis), their interactions with the environment (i.e., plastic reaction norms), their relationship with interindividual variation in traits that are influenced by many genes and contribute to disease susceptibility (i.e, intermediate quantitative traits or endophenotypes (1)) defined at different levels of hierarchical biochemical and physiological systems, and their relationship with health and disease. The goal of systems genetics is to provide an understanding of the complex relationship between the genome and disease by investigating intermediate biological processes. After investigating main effects, the first step in a systems genetics approach, as described here, is to search for gene gene (i.e, epistatic) reactions.

Alcohol addiction is a complex disease that results from a variety of genetic, social, and environmental influences. Alcoholism affected approximately 4.65 percent of the U.S. population in 2001-2002, producing severe economic, social, and medical ramifications. Researchers estimate that between 50 and 60 percent of alcoholism risk is determined by genetics. This strong genetic component has sparked numerous linkage and association studies investigating the roles of chromosomal regions and genetic variants in determining alcoholism susceptibility. To date, some of these studies have identified potential susceptibility genes. However, the complex etiology of alcoholism lends itself to further investigation that takes into account the multiple layers of interaction between genes within the context of both the genome and environment.

Systems genetics offers a new approach to studying the progression of multifaceted diseases such as alcoholism. This new and emerging field is the result of the synergy of disciplines such as bioinformatics, biotechnology, epidemiology, genetics, molecular biology, physiology, psychology, and statistics, all of which contribute to a more complete understanding of the interactions and functions of the entire genome with given ecological and sociological contexts. Detecting, characterising, and interpreting gene-gene and gene-environment interactions as risk factors for alcoholism is an important first step in a systems genetics approach that combines genomimics (2) and proteomics (3) data with methods to understand how biological processes work together to determine human health. This approach does not, however, negate the need to look for variants that directly impact disease independent of interaction effects (main effects) within the data.

A complete review of all results from genetic, genomic, proteomic, and metabolic studies of alcoholism is beyond the scope of this review. This chapter focuses on recent literature involving studies of genes selected based on biochemical evidence for their role in disease (i.e., candidate genes) and genome-wide studies, followed by an overview of the interaction among genes (i.e., epistasis) and its current and potential application in the study of alcoholism. This chapter concludes with a discussion of several methods currently being developed that incorporate a systems approach to genetics and their potential applications for the future study of alcoholism.

Alcoholism Genetics

The genetic architecture of susceptibility to a disease such as alcoholism can be defined as (1) the number of genes directly or indirectly involved, (2) the interindividual variation in those genes, and (3) the magnitude and nature of their specific genetic effects. Alcoholism develops in susceptible individuals as a result of genetic, environmental (e.g., alcohol consumption), and social influences, as well as their propensity for risk-taking bahaviours. Because of this complex etiology, multiple levels of information must be integrated to more completely understand the genetic architecture of alcoholism. In the progression of multifactorial diseases such as alcoholism, gene-gene interactions result in a variety of differentially expressed proteins. These proteins also interact, resulting in certain biochemical and physiological characteristics that, in the presence of certain environmental influences, result in alcoholism. Although studies of alcoholism's etiology have been successful in identifying a few candidate genes for susceptibility, interindividual variation in these genes accounts for only a small proportion of the overall heritability of the disease. Much of the remaining heritability is potentially due to DNA sequence variations, with effects that are dependent on contexts defined by the rest of the genome and the environment.

This chapter first review what currently is known about the role genetics plays in alcoholism and then gives a brief overview of the key findings from candidate gene and genome-wide studies. These studies confirm the role of genetics in the development of alcoholism and elucidate the need for a systems-based approach to the study of the genetic basis of the disease.

Candidate Gene Studies

Two basic strategies are used to identify genetic risk factors for common human diseases. The most common approach is to focus on a few candidate genes. This targeted approach is popular because a biological basis exists for the hypotheses being tested. Alternatively, genetic variations (i.e., polymorphisms) from across the human genome can be measured in a high-throughput manner to search for genetic risk factors without making assumptions about which genes might be important. This latter genome-wide approach is popular because much more information is examined. Researchers frequently debate the advantages and disadvantages between candidate gene and genome-wide strategies as well as the differences in genome-wide strategies themselves. Risch (2000) suggests that genome-wide association studies, which compare the genomes of people with an illness (i.e., cases) with unaffected people (i.e., controls), may be more sensitive toward finding effects for complex diseases than genome-wide linkage studies, which seek to identify regions of the genome that are associated with disease risk.

The candidate gene approach is more direct and hypothesis-based and, therefore, perhaps more likely to have significant findings, although less likely to find novel associations. The genes most extensively examined by candidate gene studies have been those involved in alcohol (i.e., ethanol) metabolism and in neurological pathways responsible for increased risk taking and "reward" stimulation from ethanol. The metabolic genes most frequently studied include those for the enzymes alcohol dehydrogenase (ADH), aldehyde dehydrogenase (ALDH), catalase, and cytochrome P450 2E1 (CYP2E1). ADH is responsible for 80 percent of ethanol's metabolism to acetaldehyde, which is then further metabolised to acetate by ALDH. CYP2E1 metabolises approximately 10 percent of ethanol and, because of its lower affinity for ethanol, is largely active only when ADH is saturated.

Researchers also have studied various genes related to the brain chemistry of alcoholism and specific chemicals (i.e., neurotransmitters) involved in addiction. Such research has examined genes for the binding sites (i.e., receptors) for the neurotransmitter gamma-aminobutyric acid (GABA); opioid receptors; components of the pathways for the neurotransmitters serotonin, dopamine, and glutamate, as well as the enzyme catechol-O-methyl-transferase (COMT), which is involved in the inactivation of dopamine; and the neurotransmitter neuropeptide Y (NPY).

Candidate gene association studies also help to focus on genetic variants that may be directly linked to pathophysiology, as reviewed by Kohnke (2007). Several genes, including those for neurotr-ansmitters such as dopamine as well as those genes mentioned above, have undergone frequent investigation in candidate association as well as linkage studies. Some have shown promising, and others conflicting, results.

The candidate gene approach has not only confirmed a genetic component of alcoholism but also has brought important understanding to disease etiology and may yield further insight when integrated with gene expression and proteomic analysis. Though the candidate approach has proven useful, genome-wide studies may provide a more comprehensive view of whole-genome interaction in the etiology of alcohol addiction.

Genome-Wide Studies

A first step toward advancing our understanding of the role of genetics in the development of alcoholism is to gather genetic data on a genome-wide scale. As previously noted, studies to date have focused on a limited number of candidate genes. Although these studies have furthered our understanding of the disease and will continue to play an important role in our understanding of alcoholism, the advent of new genomic and computational methods is making it possible to broaden our knowledge of the disease through a more inclusive whole-genome approach.

There are two main approaches to genome-wide analysis—association and linkage. Association studies examine genetic polymorphisms associated with case or control status, whereas linkage studies investigate the inheritance of specific locations on a chromosome (i.e., loci) within family lines. Though these two approaches are quite different in procedure and analysis, both are being greatly advanced by commercially available technology. The discussion of systems genetics below cites examples of both approaches.

The largest genome-wide study of alcoholism to date has been the Collaborative Study on the Genetics of Alcoholism (COGA). This study collected data from families with alcoholism and has been used for both linkage and association analyses. Researchers have identified candidate susceptibility regions on chromosomes 1, 2, and 7, including susceptibility and protective regions within the neurexin I (NRXNI) gene on chromosome 2. Another study by Namkung and colleagues (2005) was able to show that association analysis of COGA data pointed to a significant gene on chromosome 7, as well as 13 genes associated with both alcoholism and schizophrenia. The genes in these regions represent candidates for independent main effects for susceptibility to alcoholism. Some of the COGA victories include associations that have been subsequently substantiated by other investigations, such as GABA receptor alpha (GABRA2), cholinergic muscarinic 2 receptor (CHRM2), and ADH4. The challenge thereafter is to identify DNA sequence variations that influence susceptibility primarily through nonlinear interactions (i.e., the total interaction is not the sum of the influence of its interacting parts) with other genes or environmental factors. The complex interactions resulting in differential disease susceptibility and progression necessitate further investigation of epistatic interactions, which occur when the action of one gene is modified by one or several other genes.

Epistasis or Gene-Gene Interaction

Epistasis has been defined in multiple ways. The following section reviews two types of epistasis—biological and statistical—including aspects of biological epistasis, current methods used to study statistical epistasis, and analytical challenges associated with studying epistasis for genome-wide data and possible strategies for overcoming these challenges. Biological epistasis results from physical interactions among biomolecules (e.g., DNA, RNA, proteins, enzymes, etc.) and occurs at the cellular level in an individual. This type of epistasis is what Bateson (1909) had in mind when he coined the term. Statistical epistasis was first defined by Fisher (1918) as a mathematical phenomenon that occurs at the population level and is realised when there is interindividual variation in DNA sequences.

Understanding biological epistasis is one important motivation for studying statistical epistasis. With alcoholism, researchers have focused more on the direct study of biological epistasis at the cellular and biochemical level. However, a wide range of analytical tools is available for the study of statistical epistasis in human populations that could be applied to this disease. Methods for detecting statistical epistasis, described below, include linear and logistic regression, combinatorial partitioning, restricted partitioning, set association analysis, genetic programming of neural networks, symbolic discriminant analysis, and multifactor dimensionality reduction (MDR). The following section focuses first on studies of biological epistasis and then reviews some of these statistical methods.

Biological Epistasis and Alcoholism

Though gene-gene interactions are expected to play an important role in alcoholism, few studies have investigated epistasis in this disease. As mentioned above, members of the ADH gene family are common candidates for alcoholism susceptibility genes. As a model system, fruit flies (i.e., Drosophila) have been used to study epistasis in ADH genes and genes for other metabolic enzymes in relation to larval tolerance of ethanol. In one study, Freriksen and colleagues (1994) discerned differences in the metabolism of ethanol by measuring the ratios of metabolic intermediates that were "fluxed" through different branches of the ethanol metabolism pathways. They found that ethanol metabolism varied depending on the particular ADH genes present (i.e., ADH genotype). The authors suggested that the changes in intermediate ratios through the pathway might be the foundation for observed statistical epistatic interactions.

The biological epistasis of alcoholism also has been studied in reference to neurological genes. As demonstrated by Job and colleagues (2007), in a study of a type of ethanol-stimulated opioid receptor (i.e., the V-opioid receptor [MOPr]) in mice, epistatic interactions may be sexually dimorphic. The researchers found a sex-genotype interaction regarding the level of dopamine released in mice with the MOPr gene deleted (i.e., MOPr knockout mice) when they were stimulated with ethanol in the ventral striatum, with females showing a larger reduction.

Palmer and colleagues (2003) found a potential interaction between polymorphisms of the dopamine receptor D2 (DRD2) gene by constructing DRD2 knockout mice against two different genetic backgrounds (136 and 129). B6 mice previously were shown to have less stimulation in response to ethanol than 129 background mice. The two types of DRD2 knockouts showed different locomoter stimulator and locomotor sensitisation, demonstrating that there was an epistatic interaction between DRD2 and the genetic background. DRD2 also has been a popular target for studies of statistical epistasis, as described below.

Statistical Epistasis and Alcoholism

In contrast to biological epistasis, statistical epistasis is a population-level phenomenon that arises from linear and nonlinear patterns of variation in genotypes and complex traits such as alcoholism. As such, detecting and characterising statistical epistasis requires special analytical modelling methods. An association study by Osier and colleagues (2004) found a potential epistatic interaction between the ADHIB and ADH7 genes among a Han Chinese population. The ADH variant ADHI B Arg 47 His previously was found to be protective against alcoholism. This protective effect was not solely related to the ADHIB gene but to an interaction with ADH7 (or a locus that occurs with it more often than would be expected by random chance [i.e., a site in linkage disequilibrium with it]). The study included analysis of sets of closely linked genetic variants that tend to be inherited together (i.e., haplotypes) and 2-x-2 contingency tables, which are used to record and analyse the relationship between two or more variables, to discern a statistically significant, though relatively weak, protective effect of the ADH7 Styl site.

Neurological statistical epistasis studies include a study among three different Taiwanese populations that examined three different DRD2 polymorphisms. The results showed no association between the DRD2 polymorphisms and alcoholism when considered individually or as haplotypes. The minor (Al) allele of DRD2 and major (G1) allele of GABA receptor beta 3 (GABRB3), however, have been associated with alcoholism risk independently and in combination in a study of severely alcoholic and nonalcoholic Caucasians. This discrepancy regarding the statistical effect of DRD2 variations could be due to several factors, including ethnicity or an effect of DRD2 variation that is only detectable when epistasis is considered. Also, in a recent association study, COGA researchers have found that alcoholism association with the DRD2 region actually may be the result of an association with the nearby ankyrin repeat and kinase domain containing 1 (ANKKI) gene.

Chapter 5

Oxidative Stress and Mitochondrial Damage

The objectives of this study Were to determine whether differences in the size and composition of coarse (2.5-10 [micro]m), fine (< 2.5 [micro]m), and ultrafine (< 0.1 [micro]m) particulate matter (PM) are related to their uptake in macrophages and epithelial cells and their ability to induce oxidative stress. The premise for this study is the increasing awareness that various PM components induce pulmonary inflammation through the generation of oxidative stress. Coarse, fine, and ultrafine particles (UFPs) were collected by ambient particle concentrators in the Los Angeles basin in California and used to study their chemical composition in parallel with assays for generation of reactive oxygen species (ROS) and ability to induce oxidative stress in macrophages and epithelial cells.

UFPs were most potent toward inducing cellular heme oxygenase-1 (HO-1) expression and .depleting intracellular glutathione. HO-1 expression, a sensitive marker for oxidative stress, is directly correlated with the high organic carbon and polycydic aromatic hydrocarbon (PAH) content of UFPs.

The dithiothreitol (DTT) assay, a quantitative measure of in vitro ROS formation, was correlated with PAH content and HO-1 expression. UFPs also had the highest ROS activity in the DTT assay. Because the small size of UFPs allows better tissue penetration, we used electron microscopy to study subcellular localisation. UFPs and, to a lesser extent, line particles, localise in mitochondria, where they induce major structural damage. This may contribute to oxidative stress. Our studies demonstrate that the increased biological potency of UFPs is related to the content of redox cycling organic chemicals and their ability to damage mitochondria. Key words: concentrated ambient particles, dithiothreitol assay, heme oxygenase-1, mitochondrial damage, oxidative stress, polycyclic aromatic hydrocarbon, ultrafine particles.

Epidemiologic studies have shown associations between ambient air particulate matter (PM) and adverse health outcomes, including increased mortality, emergency room visits, and time lost from work and school. The underlying toxicologic mechanisms by which air pollutant particles induce adverse health effects are of intense scientific interest and have been earmarked as a key scientific priority by the National Academy of Sciences [National Research Council (NRC) 1998]. This includes a call for research on the physicochemical properties that promote particle toxicity. PM with aerodynamic diametre < 2.5 [micro]m ($P[M_{2.5}]$) is currently regulated by the U.S. Environmental Protection Agency.

Within that spectrum of particle sizes, ultrafine particles (UFPs), defined as having an aerodynamic diametre < 0.1 [micro]m, may have a central role in health effects of PM. Primary UFPs are formed during gas-to-particle conversion or during incomplete fuel combustion. Due to their small size, high number concentration, and relatively large surface area per unit mass, UFPs have unique characteristics, including increased adsorption of organic molecules and enhanced ability to penetrate cellular targets in the lung and systemic circulation.

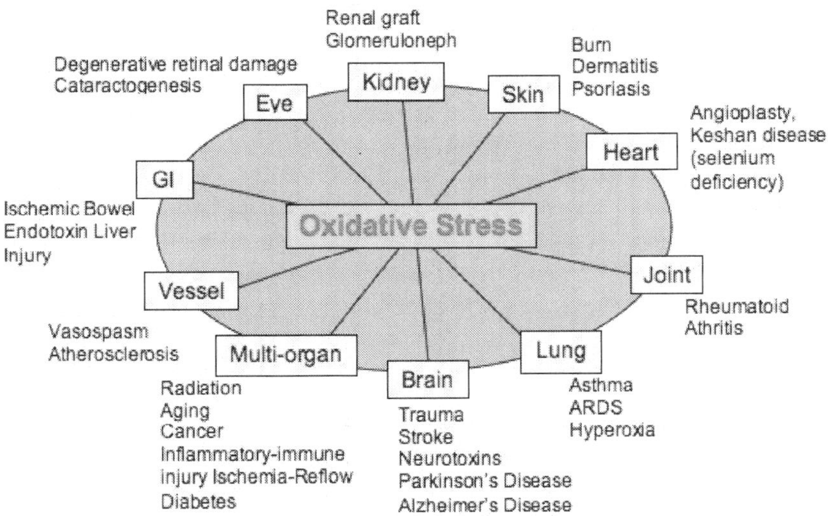

Particle composition may also be critical in PM toxicity. We are interested in organic PM compounds because organic extracts made from diesel exhaust particles (DEPs) mimic intact particles in their ability to form reactive oxygen species (ROS). One of the major advances in PM research has been the recognition that the organic and metal PM components can induce proinflammatory effects in the lung due to their ability to cause oxidative stress. Quinones present in PM can act as catalysts to produce ROS directly and may be key compounds in PM-based oxidative stress. PAHs can induce oxidative stress indirectly, through bio-transformation by cytochrome P450, expoxide hydrolase, and dihydrodiol dehydrogenase to generate redox active quinones. The involvement of quinones and PAHs was confirmed by demonstrating that compounds present in aromatic and polar fractions of DEP extracts mimic the pro-oxidative effects of intact particles in bronchial epithelial cells and macrophages. DEPs also induce cytochrome P450 1A1 induction in bronchial epithelial cells. Animal and human experiments confirm that DEPs and PAHs derived from DEPs promote allergic airway inflammation and cytochrome P450 1A1 induction in the lungs of exposed mice. Epidemiologic studies have also shown an association between PM exposure and asthma exacerbation.

The Versatile Aerosol Concentration Enrichment System (VACES), which uses three parallel sampling lines to collect concentrated ambient coarse, fine, and ultrafine particles for biological analysis, is now available for use in toxicologic studies aimed at identifying the relative toxicity of the different particle sizes. This technology enables us to probe the relationship between particle size, chemical composition, and toxicity. These concentrators are mobile and can be used to test hypotheses about particle toxicity in the Los Angeles basin in California. Concentrated air particulates (CAPs) of different sizes were collected to study their oxidative stress effects and subcellular localisation in cultured macrophages and epithelial cells. We demonstrate that UFPs are more potent than fine (< 2.5 [micro]m) or coarse (2.5-10 [micro]m) particles toward inducing oxidative stress. This effect may be explained by adsorbed chemicals (organics and metals) capable of generating ROS and the ability of UFPs to localise in mitochondria.

Ambient Particle Collection

Ambient coarse particles (2.5-10 [micro]m), fine plus ultrafine particles (< 2.5 [micro]m), and UFPs (< 0.15 [micro]m) were collected in the Los Angeles basin during November 2001-March 2002 using the VACES. Coarse particles were concentrated using a single nozzle virtual impactor, while fine and ultrafine particles were concentrated by drawing air samples through two parallel lines, using 2.5 [micro]m and 0.15 [micro]m cut-point preimpactors, respectively, to remove larger-sized particles. These particles are drawn through a saturation-condensation system that grows particles to 2-3 [micro]m droplets, which are subsequently concentrated by virtual impaction. Highly concentrated particle suspensions were obtained by connecting the VACES output to a sterilised liquid impinger. Aerosols were collected using ultrapure deionised water (resistivity 18.2 megaohm; total organic compounds < 10 ppb; particle-free; bacteria < 1 colony forming unit/mL) as the collection medium.

The concentration enrichment process does not alter the physical, chemical, and morphologic properties of the particles. We determined the total amount of particulate loading in the collection medium by multiplying the ambient concentration of each PM mode by the total air sample volume collected by each VACES line. The particle concentration in the aqueous medium was then calculated by dividing the particle loading by the total volume collected in that time period. Five sample sets were collected, two at the University of Southern California (USC), and three at Claremont. USC is a typical urban site located 3 km south of downtown Los Angeles. This is a site in which aerosols are mostly generated from fresh vehicular emissions. Claremont is a receptor site approximately 45 km east (i.e., downwind) of downtown Los Angeles. In that location, ambient PM originates mostly from advection of polluted air parcels originally emitted in urban Los Angeles, after "aging" in the atmosphere for a few hours, as well as from secondary photochemical processes.

Particle Chemical Analysis

Samples were collected on Teflon and quartz filters with a Micro Orifice Uniform Deposit Impactor (MOUDI; MSP Corporation, Shoreview, MN) for chemical analysis. We used Teflon filters to determine the metal and trace element content by X-ray fluorescence and quartz filters to determine the organic carbon (Mn[O_2]-catalysed C[O_2] formation), sulfate (ion chromatography), and nitrate (ion chromatography) contents. PAH content for each CAPs set was determined by an HPLC-fluorescence method that detects a signature group of 16 PAHs.

Cellular Stimulation and Heme Oxygenase 1 (HO-1) Immunoblotting

We used two cell lines in the study: RAW 264.7 and BEAS-2B. RAW 264.7 is a murine macrophage cell line that mimics the oxidative stress response of pulmonary alveolar macrophages in response in DEP exposure. BEAS-2B is a transformed human bronchial epithelial cell line, which mimics the oxidative stress response of primary bronchial epithelial cells. For RAW 264.7 culture, particle suspensions were reconstituted with Dulbecco's Modified Eagle's Medium powder, a culture medium component that rapidly dissolves in deionised water. This culture medium was further replenished with 10% fetal calf serum and a 1:200 dilution of penicillin/streptomycin/amphotericin B. For BEAS-2B cells, particle suspensions were made up in hormonally defined F12 medium. After incubating cells for 16 hr, we used 100 [micro]g of lysate protein for HO-1 immunoblotting. Densitometric analysis was performed on a laser Personal Densitometre SI using ImageQuant software (both from Amersham Biosciences, Piscataway, NJ).

GSH/GSSG assay

Total glutathione and oxidised glutathione (GSSG) were measured in a glutathione reductase recycling assay. We calculated the amount of total glutathione and GSSG in the samples from the standard curves. The amount of reduced glutathione (GSH) was calculated by subtracting the amount of GSSG from that of the total glutathione.

DTT assay

The dithiothreitol (DTT) assay quantitatively measures the formation of ROS by quinone catalysis. In the presence of quinones, 1 mol DTT + 2 mol [O_2] generate 1 mol DTT-disulfide + 2[O_2^-]

a: Q + DTT $\rightarrow$ semi-Q + DTT-thiyl

b: Q + DTT-thiyl $\rightarrow$ semi-Q + DTT-disulfide

c: 2 semi-Q + 2[O_2] $\rightarrow$ 2Q [+ or -] 2[O_2^-]

Net reaction: DTT + 2[O_2] $\rightarrow$ DTT-disulfide + 2[O_2^-]

The loss of DTT is followed by its reaction with 5,5'-dithiobis-(2-nitrobenzoic acid) (DTNB), which is converted to 5-mercapto-2-nitrobenzoic. We incubated the PM sample (5-50 [micro]g/mL with 10 [micro]M DTT in a Tris buffer at pH 8.9 for 10-90 min. Aliquots of the incubation mixture were transfered to the DNTB solution and the optical density read at 412 nm.

Electron Microscopy

We performed electron microscopy as previously described. Thin sections were cut with a Reichert-Jung ultracut and ultramicrotome. Copper grids were stained with lead citrate and uranyl acetate and photographed in a Hitachi electron microscope. Particulate organic carbon and PAH content.

To determine whether there is a link between UFP composition and biological effects, CAPs were collected at two outdoor sites in the Los Angeles basin with the VACES. Chemical analysis of the CAPs indicate that UFPs have a significantly higher organic ($p < 0.01$) and elemental carbon (($p < 0.001$) content than fine plus ultrafine (designated "fine") or coarse particles. Coarse and fine particles had a higher metal content than UFPs. PAH content for each set of CAPs was determined and averaged for both collection sites; there was a statistically significant difference in total PAH content in UFPs compared to fine ($p = 0.04$) and coarse ($p = 0.03$) PM. The PAH content of UFPs at USC was significantly higher than the PAH content at Claremont, which reflects the fact that particles collected at the source site (USC) are derived from primary emissions that are far more abundant in the urban areas of Los Angeles than in receptor areas.

The DTT Assay of ROS Formation by Particles of Varying Size

The ability of PM to generate ROS was assessed with the DTT assay. Quinones with appropriate redox potentials can transfer electrons from DTT to oxygen. We used this reaction to determine the ability of PM to generate ROS in vitro. The DTT assay demonstrated that UFPs had significantly higher redox activity than fine and coarse PM. Averaging of the data revealed that the redox cycling capacity of UFPs was 21.7—and 8.6-fold greater than coarse and fine PM, respectively. Regression analysis of the DTT assay and PAH content showed a correlation coefficient ([r^2]) of 0.98, suggesting the electron transfer capacity of CAPs is consistent with their organic chemical content.

Particle Chemical Composition and Oxidative Stress

Quinones and other redox-active compounds present in PM generate ROS and oxidative stress. We have demonstrated that DEP-induced oxidative stress generates hierarchical effects in pulmonary alveolar macrophages and bronchial epithelial cells. Low levels of oxidative stress activate antioxidant defences, whereas higher levels of oxidative stress lead to proinflammatory and cytotoxic effects. An example of an antioxidant response is HO-1 expression via the antioxidant response element in its promoter. Utilising an immunoblotting technique to assess HO-1 expression in RAW 264.7 cells, UFPs were more potent than fine or coarse particles. Densitometric analysis demonstrated significantly higher HO-1 expression in ultrafine over fine ($p = 0.001$) and coarse ($p = 0.001$) particles, respectively. The increased potency of UFPs was seen for all CAPs collections.

Sufficient CAPs were collected in March 2002 to study HO-1 expression in the bronchial epithelial cell line, BEAS-2B, in parallel with RAW 264.7. The BEAS-2B response mimics the DEP-induced oxidative stress response in human bronchial epithelial cells. Immunoblot analysis shows that UFPs but not coarse or fine particles induce HO-1 expression in BEAS-2B cells. To demonstrate that these effects reflect differences in the level of oxidative stress, we compared HO-1 expression to changes in the reduced (GSH) to oxidised (GSSG) glutathione ratio. These data show abundant HO-1 expression by UFPs or 50 [micro]g/mL fine PM, which is accompanied by a sizable drop in glutathione ratios. In contrast, coarse particles had no effect on either biological response. Regression analysis showed a correlation coefficient ([r^2]) of 0.97 between HO-1 expression and the DTT assay. Overall, there is a strong correlation between particle size, chemical composition, ROS-generating capacity, and cellular oxidative stress.

UFP Localisation and Mitochondrial Damage

In defining the mechanistic features of PM toxicity, a key question is the subcellular localisation of PM. This may determine ROS generation, as demonstrated by [O_2^-] generation in lung microsomes during incubation with DEP extracts. Subcellular DEP targets include mitochondria, as demonstrated by the ability of organic DEP extracts to induce structural mitochondrial damage.

After exposure to CAPs, there were clear differences in the ultramicroscopic features of RAW 264.7 cells exposed to different particle sizes. Whereas coarse particles collected in large cytoplasmic vacuoles, UFPs frequently lodged inside mitochondria. Mitochondrial architecture remained intact in coarse PM incubations, but cells incubated with UFPs showed extensive disruption of mitochondrial cristae, resulting in a vacuolar cellular appearance. These changes were time dependent, with fewer particles collecting inside mitochondria during shorter incubations (not shown). In cells exposed to fine particulates (which includes some UFPs), some particles lodged inside mitochondria but did not show the same degree of ultrastructural damage.

Electron microscopy showed similar features in BEAS-2B cells—namely, considerable mitochondrial damage by UFPs, resulting in the formation of concentric structures, known as myelin figures. These structures result from the disassociation of lipoproteins, which facilitates water uptake and intercalation between lamellar membrane stacks. Similar to RAW 264.7 cells, UFPs lodged inside damaged mitochondria. Cells incubated with coarse or fine particles showed lesser mitochondrial damage. The extent of mitochondrial damage is in accordance with the redox cycling potential of the particles, as well as the HO-1 and glutathione results.

Our data demonstrate that the UFP mode in the Los Angeles basin is more potent than fine and coarse PM toward inducing oxidative stress as measured by the DTT, HO-1, and glutathione assays. Electron microscopy also indicates subcellular penetration and mitochondrial damage by UFPs and, to a lesser extent, fine particles. The findings correlate with PM organic carbon and PAH composition, suggesting a role of organic agents in generating redox activity. The results from the DTT assay indicate UFPs are capable of producing greater ROS on a microgram basis than fine and coarse particles. This is the first time that a quantitative assay has been used to directly measure ROS generation by CAPs. Kumagai et al. (2002) focused the assay on assessment of quinones, whereas we have applied the assay to UFPs and fine and coarse particles. The DTT assay provides a quantitative measure of the relative redox activity of different PM sizes in the Los Angeles basin.

UFPs contain a higher percentage of organic carbon than fine and coarse particles, and this has relevance to the biologic potency of these particles. The enhanced biologic potency of UFPs is directly correlated with the PAH content. Although PAHs are capable of inducing ROS production in macrophages, it is also possible that these compounds may be a surrogate for other redox cycling chemicals in the DTT assay. We do not exclude a contribution by transition metals, which may interact with organic PM components in ROS generation. These data are in accordance with the growing awareness that oxidative stress plays a key role in the induction of airway inflammation. Recently, we demonstrated that macrophages and epithelial cells exhibit a stratified oxidative stress response to increasing concentrations of DEPs.

The stratified response commences with HO-1 expression when the GSH/GSSG ratio is minimally disturbed, proceeds to Jun kinase activation at intermediary levels of oxidative stress, and culminates in cellular toxicity at high oxidative stress levels. Ambient CAPs mimic the effects of organic DEP extracts, with UFPs showing increased potency in depressing the cellular GSH/GSSG ratio. The significance of Jun kinase activation is the transcriptional activation of cytokine, chemokine, and adhesion receptor promoters. These products play a role in the pro-inflammatory effects of PM in the lung and possibly also the cardiovascular system. The finding of a significant correlation between heme oxygenase activity, GSH/GSSG ratio, and redox activity as measured by DTT production provides further evidence for the role of ROS generation in PM toxicity.

The biological significance of HO-1 expression in the lung is the antioxidant effect of its catabolic product, bilirubin. In the process of heme catabolism, HO-1 also generates a gaseous substance, CO, which exerts anti-inflammatory effects in the lung and is exhaled in the expired air. It is interesting, therefore, that in a study in which normal human volunteers were exposed to DEPs, CO levels in the expired air was a more sensitive exposure marker than the presence of inflammatory products in the bronchoalveolar fluid. This is in agreement with the exquisite sensitivity of the HO-1 promoter to oxidative stress in vivo and in vitro. HO-1 expression and CO generation are markers for airway inflammation in asthma. Monitoring of CO levels in the expired air may be a useful marker for evaluating the pro-oxidative and proinflammatory effects of CAPs in the respiratory tract.

How exactly UFPs gain access to and induce mitochondrial damage is unknown. One possibility is that ROS generated outside of the mitochondrion may damage this organelle, allowing access to the particles. This is compatible with the ability of organic DEP extracts to induce ultrastructural mitochondrial damage in the absence of particles. Our previous studies have demonstrated that organic DEP chemicals induced pro-apoptotic effects in macrophages and bronchial epithelial cells.

This effect may be mediated through the perturbation of mitochondrial permeability transition pore, which sets in motion cytochrome c release, caspase activation, and superoxide production in the mitochondrial inner membrane. Ultramicroscopic visualisation of human macrophages and BEAS-2B cells incubated with organic

DEP extracts showed that the appearance of apoptotic bodies were accompanied by changes in mitochondrial morphology, including mitochondrial swelling and a loss of cristae. Another possibility is that UFPs gain access to mitochondria because of their small sizes.

These particles might then release redox cycling chemicals that damage the inner membrane. All considered, we propose that enhanced tissue penetrance and ability to generate oxidative stress render UFPs more damaging at cellular level and consequently contribute to the adverse health effects of UFPs in the Los Angeles basin.

These findings may be of importance for PM regulation. Currently, the manufacture of cleaner combustion engines relies on mass output standards but do not consider the output of large numbers of UFPs, which have very low mass. Our data show that UFPs are more potent than $P[M_{2.5}]$ and $P[M_{10}]$ that contribute the majority of mass in the HO-1 and DTT assays. It may be necessary to consider standards based on particle number instead of mass if further studies confirm the differential toxicity of UFPs. Further research to more fully characterise the toxicity of UFPs in relation to particle number, surface area, and chemical composition is needed.

C-type Cytochrome Assembly in Saccharomyces Cerevisiae

The c-type cytochromes, also referred to as cytochrome c, represent a universal class of hemecontaining proteins that function as electron carriers in the energy-transducing pathways of bacteria, plastids, andmitochondria. Because cytochromes c carry a heme covalently attached to a CXXCH motif, they constitute an attractive object of study to address the question of cofactor protein assembly. The biochemical requirements for cytochrome c assembly were deduced from in vivo and in vitro studies, and the conclusion is that both apocytochromes c and heme are transported independently across at least one biological membrane and maintained as reduced prior to catalysis of the heme attachment reaction. Bacterial cytochromes c are assembled in the periplasmic space, a compartment where cysteine pairs in proteins form disulfide bonds in reactions catalysed by dedicated enzymes. The current thinking holds that a c-type apocytochrome is a substrate of the disulfide bond-forming pathway, which introduces an intramolecular disulfide between the two cysteines of the CXXCH sequence. This disulfide needs to be reduced to a dithiol to provide free sulfhydryls for the heme ligation. Consistent with this view is the fact that groups of specific oxido-reductases that constitute a transmembrane dithiol-disulfide relay from the cytosol to the periplasmic space have been shown to function as c-type cytochrome assembly factors. The proposal that the components of this pathway control the in vivo redox status of the CXXCH sulfhydryls has been inferred from the presence of motifs in their protein sequences that are consistent with a function in redox chemistry and also from the demonstration that their recombinant forms participate in dithiol-disulfide exchange reactions. Moreover, the ability of exogenous thiol compounds to bypass the lack of these factors in vivo substantiates the view that the redox components have a disulfide-reducing activity in the pathway.

While the role of these pathways is well established in bacteria, much less is known about the components that catalyse thiol/disulfide chemistry in the mitochondrial intermembrane space (IMS), which is topologically equivalent to the bacterial periplasm. By analogy with the bacterial pathways, the participation of redox-active factors that catalyse thiol formation is expected, as the mitochondrial IMS houses two c-type cytochromes, the soluble cytochrome c and the membrane-bound cytochrome c_1, both of which function in respiration. In fungi, heme attachment to apocytochromes c and c_1 is dependent upon the IMS resident cytochrome c and c_1 heme lyases, CCHL and CC_1HL, although the exact role of these lyases in the assembly process is still unclear. Conversion of apocytochrome to holocytochrome c depends only on CCHL, while apocytochrome c_1 can be acted upon by both CCHL and CC_1HL. In animals, apoforms of cytochromes c and c_1 are assembled by a unique heme lyase, HCCS, which carries both the CCHL and CC_1HL activities.

Cyc2p, a component first described as a mitochondrial biogenesis factor in yeast, was recently rediscovered in the context of cytochrome c_1 maturation. Cyc2p is located at the mitochondrial inner membrane with its C-terminal domain containing a non-covalently bound FAD exposed to the IMS. A redox function for Cyc2p is likely based on the finding that a recombinant form of the molecule exhibits a NAD(P)Hdependent reductase activity. However, as Cyc2p activity is not essential for the maturation process, a functional redundancy was postulated based on the fact that a cyc2-null mutant still assembles holoforms of cytochromes c and c_1. The absolute requirement of Cyc2p was revealed via genetic analysis of the cyc2-null cyt1-34 combination that displays a synthetic respiratory-deficient phenotype with loss of holocytochrome c1 assembly. The cyt1-34 mutation maps to the gene encoding cytochrome c_1 and results in a CAPCH heme-binding site replacing the wild-type CAACH site. The synthetic interaction is specific for the cyt1-34 allele carrying the A-to-P mutation and is not observed in a cyc2-null cyt1-48 strain carrying an A-to-D mutation at the heme-binding site of apocytochrome c_1. The fact that Cyc2p becomes essential when the cytochrome c_1 hemebinding site carries an A-to-P mutation suggests that the CXXCH motif could be the target of Cyc2p action in vivo. One possible interpretation for this observation is that the P residue alters the reactivity of the cysteinyl thiols to redox chemistry so that the apocytochrome c_1 CAPCH heme-binding site occurs in an oxidised (disulfide) form, which must be reduced in a Cyc2p-dependent reaction before heme attachment can occur.

In this chapter, we have undertaken a genetic approach to elucidate this pathway and searched for suppressors that alleviate the respiratory deficiency of the cyc2-null cyt1-34 strain. Either overexpression of CC_1HL or replacement of the P mutation in the heme-binding site by H, L, S, or T residues restore the assembly of holocytochrome c_1. In vitro measurement of redox potential of apoforms of CA(A/P/T)CH cytochrome c1 indicates that there is no change in the thermodynamic stability of the disulfide at the CXXCH motif that could account for the Cyc2p-dependent assembly of cytochrome c_1. Genetic studies reveal that the replacement of the second A residue at the CAACH motif by H, L, P, S, and T residues is key in determining the conversion of apocytochrome c_1 to its corresponding holoform via the CCHL and/or CC_1HL-dependent pathway.

We also demonstrate that Cyc2p is a component dedicated to the CCHL pathway and is not required for the CC_1HL dependent assembly of cytochrome c_1. We propose that the CAPCH cytochrome c_1 is strictly dependent upon CCHL and Cyc2p for its assembly but becomes a substrate of CC_1HL upon overexpression of CC_1HL or in the presence of H, L, S, or T mutations. Construction, manipulation, and growth of yeast strains: YDM1 and YDM2 strains carry the cyt1Tkan allele and originate from the parental SMY1 (MATa) and YCT1-11C (MATa) 94cc1hl strains, respectively. PCR amplification of the cyt1Tkan cassette was performed using oligonucleotide sequences (5'-AACTGGATCCATAGACTATCTAAG-3' and 5'- GACACTATTG-AAGTGAGACG-3') and the genomic DNA of the cyt1Tkan strain from the ResGen knockout collection. Yeast cells were transformed by the lithium acetate procedure or by the one-step technique. Genetic crosses were performed as described in Dujardin et al. (1980). Media used for Saccharomyces cerevisiae contain glucose or galactose as fermentable substrates and glycerol, ethanol, ethanol/glycerol, or lactate as respiratory substrates and were described elsewhere.

Isolation of multicopy and genetic suppressors: The multicopy wild-type genomic libraries constructed in the URA3-based pFL44L vector (a generous gift from F. Lacroute) and in the LEU2-based Yep24 vector (a generous gift from M. Carlson) were used to search for suppressor genes able to alleviate the respiratory deficiency of the cyc2-null cyt1-34 strain. Transformants were plated on medium lacking uracil or leucine, replicated on lactate medium to select for respiratory proficiency, and incubated for 15 days at 28°. Plasmids retrieved through the multicopy suppressor screen were extracted from yeast transformants and propagated in Escherichia coli strains. Independent spontaneous suppressors bypassing the respiratory deficiency of the 94cyc2 cyt1-34 strain were isolated on lactate medium. Cells were grown to stationary phase and plated on respiratory medium, and suppressed strains were isolated after 10-15 days incubation at 28°.

Overexpression and purification of apocytochromes c_1: The CYT1 gene encoding the soluble form (amino acids 62- 263) of wild-type apocytochrome c1 was amplified by PCR from yeast genomic DNA and cloned into the NheI/XhoI sites of the hexahistidinyl tag plasmid pET-24b(+) (Novagen). The cysteines (C188 and C258) were replaced by serines by sitedirected mutagenesis. Mutations in the heme-binding motif (CAPCH and CATCH) were also engineered by site-directed mutagenesis. Apocytochromes c_1 were overexpressed in E. coli BL21 (DE3) strain in the presence of 1 mm isopropyl-b-dthiogalact-opyranoside for 5 hr at 30°. All recombinant apocytochromes c1 (wild-type and mutants) carry a C-terminal (His)6-tag, and as a result of the C188,258S substitutions, the only cysteines present are the two associated with the hemebinding site. Purification of His-tagged proteins was performed under denaturing conditions. Cells were harvested, resuspended in a HEPES buffer (100 mm HEPES, 100 mm NaCl, 0.01% Tween 20, pH 8.0) containing 8 m urea and 50 mm imidazole and incubated at room temperature for 1 hr. The lysate was clarified by centrifugation at 10,000 × g for 20 min at room temperature. The supernatant was then applied to the Ni-NTA resin (Qiagen). Protein renaturation and refolding were carried out directly on the column by gradient buffer exchange to eliminate urea. The resin was washed with the HEPES buffer containing 100 mm imidazole, and the proteins were eluted with 250 mm imidazole. Samples were then dialysed to remove imidazole and concentrated by using a Centriprep filter unit (Millipore). Purity was assessed by SDS-PAGE, and the concentration of the purified protein was measured by recording the absorbance at 260 nm.

Measurement of midpoint redox potentials of apocytochromes c_1: Redox titrations of disulfide/dithiol couples were performed using thiol labelling with 4-acetamido-4'- maleimidylstilbene-2,2'-disulfonic acid (AMS; Invitrogen). Wild-type and CAP/TCH apocytochromes c_1 recombinant proteins (5 mg) were incubated separately under anaerobic conditions at 25° for 5 or 16 hr in 100 mm HEPES (pH 7.0) and various concentrations of reduced and oxidised DTT at a total DTT concentration of 5 mm. Ambient potential (E_h) values for different DTT_{red}/DTT_{ox} ratios were calculated from the two-electron Nernst equation, using the literature value for the midpoint potential of DTT at pH 7.0. Incubations in the presence of 50 mm of $CuCl_2$ or 5 mm reduced DTT served to produce fully oxidised or fully reduced protein controls, respectively.

After incubation, samples were treated with 10% trichloroacetic acid (TCA) and allowed to stand on ice for 30 min. Protein precipitates were washed with 1% TCA and then with ice-cold acetone. Samples were resuspended in buffer containing 50 mm HEPES (pH 7.0), 2.5% SDS, and 10 mm AMS and incubated at 37° for 1 hr in an argon atmosphere. Reduced (AMS-labelled) and oxidised (nonlabelled) forms of apocytochromes c_1 were separated by nonreducing SDS-PAGE on a 16% gel, stained with silver nitrate, and quantified using Image J software (National Institutes of Health). Data were plotted and fitted to the Nernst equation for a single redox couple by iteration, using the solver of Excel (Microsoft). The midpoint potential (E_m) was defined as a variable, and its value was taken after the target cell, containing the average quadratic variation between the theoretical curve and the experimental data, converged to its minimum value. Best fits were obtained for an n-value of 2 (i.e., for a two-electron redox reaction). All E_m values reported correspond to the average of three independent titrations.

Mitochondrial protein preparation and analysis: Mitochondria were purified from yeast strains grown in galactose medium as previously described, and the protein concentration was measured using the Bradford reagent (Bio-Rad). For heme staining of mitochondrial c-type cytochromes, protein samples were reduced with

50 mm DTT on ice for 1 hr and separated at 4° by lithium dodecyl sulfate (LDS)-PAGE. Detection of holocytochromes c and c_1 was performed on polyvinylidene difluoride membrane (PVDF; 0.45 mm) by the enhanced chemiluminescence (ECL) method using the heme-associated peroxidase activity of holocytochromes c. For immunoblotting, mitochondrial proteins were separated by SDS-PAGE and subsequently immobilised by electro-transfer to PVDF membranes. Polyclonal antibodies raised against cytochrome c, cytochrome c1, CCHL, CC_1HL, Cyc2p, and Coq1 were used for immunodetection. Bound antibodies were detected by horseradish peroxidase-conjugated secondary antibody (Bio-Rad) and ECL reagent.

A multicopy suppressor screen of the Dcyc2 cyt1- CAPCH mutant identifies CC^sub 1^HL: For convenience, the cyc2-null cyt1-34 mutant will be referred to as 94cyc2 cyt1-CAPCH throughout the entire manuscript. To gain further insight into the function of Cyc2p, we undertook a search for multicopy suppressors able to alleviate the respiratory deficiency of the 94cyc2 cyt1-CAPCH mutant. The respiratory-deficient phenotype of the 94cyc2 cyt1-CAPCH strain is tight and characterised by a dual deficiency in holoforms of cytochrome c and c_1. Of 500,000 primary transformants, 87 respiratory-competent colonies were selected after a 10- day incubation on medium containing a respiratory substrate. Extended incubation gave rise to colonies in which respiratory proficiency was independent of the presence of the multicopy plasmid and probably caused by suppressor mutations. Diagnostic PCR amplification using CYC2- and CYT1-specific primers showed that of the 87 transformants, 44 carried a plasmid containing the CYC2 gene, whereas 36 harboured a plasmid with the cytochrome c_1-encoding gene. Both genes were expected to be recovered, as the 94cyc2 cyt1-CAPCH respiratory-deficient phenotype is synthetic.

The remaining 7 transformants displayed a slower growth phenotype, and PCR amplification failed to reveal the presence of CYC2 or CYT1 genes in the transforming plasmids. Sequencing of the insert ends in one multicopy plasmid and subsequent diagnostic PCR revealed that all slow-growing transformants carry the gene encoding CC_1HL. Spectral analysis showed that overexpression of CC_1HL is able to partially restore holocytochromes c and c_1 assembly. This result is consistent with the restoration of the respiratory growth as both cytochromes c and c_1 are required for respiration. As expected, overexpression of CYT1 in 94cyc2 cyt1-CAPCH resulted in levels of c-type cytochromes identical to that of a 94cyc2 strain while overexpression of CYC2 yielded a spectral phenotype identical to the one of the cyt1-CAPCH mutant. We also demonstrated that overexpression of either CCHL- or apocytochrome c-encoding genes in 94cyc2 cyt1-CAPCH does not restore cytochrome c or c_1 assembly. We concluded that CC_1HL acts as a multicopy suppressor that can partially compensate for the dual deficiency of the holocytochromes c and c_1 in the 94cyc2 cyt1-CAPCH mutant.

Suppressor mutations of the 94cyc2 cyt1-CAPCH mutant map to the cytochrome c^sub 1^ heme-binding site: We observed that 94cyc2 cyt1-CAPCH could revert to respiratory proficiency, allowing us to isolate SSP (for Suppressor of Synthetic Phenotype) strains. Spontaneous independent suppressors (SSP1-SSP14) of the 94cyc2 cyt1-CAPCH were isolated and further characterised. All SSP strains were restored to the same extent of respiratory growth. Genetic analysis indicated that the suppressor mutation is monogenic and linked to the original mutations. Because CYC2 and CYT1 are genetically linked and the Dcyc2 allele corresponds to a complete deletion of the CYC2 gene, we reasoned that the suppressor mutation mapped to the cytochrome c1- encoding gene (CYT1). Sequencing of the CYT1 gene in the different SSP strains showed that a single-nucleotide change resulted in the alteration of the proline in the CAPCH heme-binding site to A, H, L, S, or T residues. Note that SSP13 and SSP14 have reverted to a wild-type heme-binding site (CAACH), an expected finding as assembly of wild-type cytochrome c1 does not require Cyc2p. Analysis of the holoforms of c-type cytochromes of representative SSP strains carrying the CAHCH, CALCH, CASCH, and CATCH heme-binding sites reveals that holocytochrome c_1 accumulation is restored to the wild-type level while the level of holocytochrome c is similar to that of the 94cyc2 mutant.

The suppressor mutations do not affect cytochrome c1 assembly: The observation that the A-to-P mutation results in a slow respiratory growth associated with a decrease in holocytochrome c1 accumulation prompted us to assess the impact of the cyt1-CAHCH, -CALCH, -CASCH, and -CATCH alleles in an otherwise wild-type background. Strains carrying cytochrome c1 with a CAHCH, CALCH, CASCH, or CATCH heme-binding site were constructed by genetic crosses and examined for respiratory competence and holocytochrome c assembly. All strains were found to have respiratory growth rates similar to that of the wild type. Spectral and heme stain analyses showed that they were indistinguishable from wild type as far as holocytochrome c and

c_1 assembly is concerned. We concluded that, unlike the A-to- P mutation, the A-to -H, -L, -S, and -T substitutions in the heme-binding site have no impact on holocytochrome c_1 assembly.

Redox titrations of apocytochromes c_1 do not indicate a change in the midpoint potential of cysteines at the heme-binding site: One possible scenario to account for the fact that CAPCH cytochrome c_1 assembly is dependent upon Cyc2p is that the A-to-P mutation alters the redox chemistry of the cysteinyl thiols so that the apocytochrome c_1 heme-binding site is present in a predominantly oxidised, disulfide form. In such a scenario, the cysteine thiols necessary for heme attachment are either not present or present only to a limited extent. To assess whether mutations in the hemebinding motif of cytochrome c_1 might affect the redox chemistry of the cysteinyl thiols of the heme-binding motif in a manner that would make the oxidised disulfide state thermodynamically more stable, we chose to measure the midpoint redox potential (E_m) of recombinant wild-type (CAACH) and mutant (CAPCH and CATCH) apocytochromes c_1.

This is the best quantitative measure of the thermodynamic propensity of a cysteine pair to be either in a reduced or in an oxidised state. It originally appeared likely that CAPCH apocytochrome c_1 might be characterised by a midpoint potential value that was substantially more negative than that of the wild-type form, given our prediction that the A-to-P mutation produces a higher level of the oxidised form of the apoprotein. To test this hypothesis, redox titrations were carried out using recombinant forms of the apocytochrome c_1 that corresponded to the soluble domain of the protein and did not contain cysteine residues other than the pair present at the heme-binding site. Measurements of redox potentials were carried out using DTT_{red}/DTT_{ox} buffers to poise samples at defined ambient potential (E_h) values, prior to their treatment with the thiol modifier AMS. AMS increases the molecular weight of a protein by 0.5 kDa per thiol modified, allowing the separation of the doubly modified, reduced form from the unmodified oxidised form by electrophoresis on a nonreducing SDS gel. All titrations gave excellent fits to the Nernst equation for a single two-electron redox component, with E_m values of -266 ± 4 mV, -261 ± 3 mV, and -257 ± 3 mV for the apocytochromes c_1 carrying the CAACH (wild type), CATCH, and CAPCH variations of the heme-binding motif, respectively. Because there was little difference between the midpoint potentials determined for these three forms of the apocytochrome c_1, we concluded that the A, P, and T residues did not significantly impact the thermodynamic propensity of the heme-binding site to occur in a disulfide vs. dithiol form.

The CASCH mutation suppresses the absence of CC_1HL: The cyt1-CAPCH allele and the cyt1-CADCH mutation were originally isolated in a screen for mutations enhancing the activity of CCHL toward apocytochrome c_1 in the absence of its cognate lyase CC_1HL. To test whether H, L, S, and T mutations increase the activity of CCHL toward apocytochrome c_1, we generated the 94cc1hl cyt1-CAHCH, -CALCH, -CASCH, and -CATCH strains. In such strains, CC_1HL is absent, and apocytochrome c_1 conversion to its corresponding holoform is completely dependent upon CCHL. While cytochrome c1 with a CASCH hemebinding domain could restore the respiratory growth in a 94cc1hl strain, cytochrome c_1 with a CAHCH, CALCH, or CATCH heme-binding domain further diminished the ability of the 94cc1hl strain to grow on a respiratory substrate. Consistent with the respiratory growth phenotype, the 94cc1hl cyt1-CASCH strain exhibits enhanced level of holocytochrome c_1 while the 94cc1hl cyt1-CATCH strain accumulates lower level of holocytochrome c_1 than the 94cc1hl strain. Note that the 94cc1hl strain displays a reduced level of respiratory growth that is attributed to the weak activity of CCHL toward apocytochrome c_1. We concluded that the S mutation in the heme-binding site increased the activity of CCHL toward apocytochrome c_1 while the H, L, and T residues decreased the activity of CCHL toward its noncognate substrate.

Cyc2p is not required for CC_1HL-dependent assembly of cytochrome c_1: Earlier studies have established that the assembly of cytochrome c1 proceeds normally in the absence of CCHL, an indication that holocytochrome c_1 is synthesized independently via its cognate heme lyase, CC_1HL. To assess whether Cyc2p function was also required for the assembly of cytochrome c_1 via CC_1HL, we constructed a cchl-null cyc2-null mutant and monitored the level of c-type cytochromes by spectra. As expected, the cchl-null cyc2-null strain is deficient in holocytochrome c due to the lack of its cognate heme lyase but retains a wild-type level of cytochrome c_1, suggesting that CC_1HL does not require the activity of Cyc2p for conversion of apocytochrome c_1 to its corresponding holoform.

The function of Cyc2p is clearly redundant based on the fact that a cyc2-null mutant is not completely deficient in cytochrome c. Functional redundancy does not involve Mcr1p and Cbr1p, two proteins with mitochondrial localisation that exhibit similarity to Cyc2p.

Previous studies have identified Cyc2p, a pyridine nucleotide-dependent flavoprotein, as a redox component controlling the maturation of mitochondrial c-type cytochromes. Genetic analyses revealed that Cyc2p is not absolutely required for c-type cytochrome maturation but becomes essential when the cytochrome c_1 heme-binding site is modified from CAACH to CAPCH, an indication that the CXXCH motif could be the relevant target of Cyc2p action in vivo. Our initial interpretation of this observation was that the A-to-P mutation altered the redox chemistry of the cysteinyl thiols so that the apocytochrome c_1 CAPCH heme-binding site exists largely in an oxidised form (i.e., the cysteines form an intramolecular disulfide bond) and is no longer able to participate in the heme attachment without Cyc2p intervention.

It is now well recognised that the nature of the residues between the cysteines at a CXXC redox motif influences the propensity of this motif to be in the oxidised (disulfide) or reduced form (dithiol). Examples come from the documented changes in midpoint potential caused by mutations of the residues between the two cysteines at the redox active site of thioredoxin, protein disulfide isomerase and DsbA oxido-reductase. To better understand why Cyc2p is required for the assembly of cytochrome c_1 with a CAPCH hemebinding site, we isolated secondary mutations that enable Cyc2p-independent assembly of cytochrome c_1. All the bypass mutations that we isolated change the proline at the cytochrome c_1 heme-binding motif to A (wild-type), H, L, S, or T residues and restored a wildtype level of holocytochrome c_1 assembly. As stated above, we had originally interpreted this result in the context of a change in thiol chemistry at the hemebinding site and postulated that cytochrome c_1 with a CAHCH, CALCH, CASCH, or CATCH is able to be assembled because the presence of a H, L, S, and T residue instead of P results in a shift of the equilibrium of the CXXCH heme-binding site toward the reduced form (i.e., dithiol) that is competent for the heme ligation reaction.

Measurements of midpoint potentials reveal that there is no significant difference between wild-type, CAPCH, and CATCH cytochromes c_1, anindication that, at redox equilibrium, the thermodynamic stability of the disulfide bond at the heme-binding site is not influenced by the P and T mutations. Hence it is unlikely that the Cyc2p-dependent assembly of CAPCH cytochrome c_1 can be explained by a change in the equilibrium dithiol-disulfide ratio of the apocytochrome c_1 heme-binding site. However, because the midpoint potential reflects only the redox state at thermodynamic equilibrium, it is conceivable that the P residue has altered the reactivity of cysteines in the CXXCH motif so that Cyc2p activity is required for the heme attachment to apocytochrome c_1.

We had previously shown that apocytochrome c_1 is a substrate of both CC_1HL and CCHL. While we demonstrate in this work that Cyc2p is not required for the CC_1HL-dependent assembly of apocytochrome c_1, its activity is absolutely essential for the CCHL-dependent assembly of apocytochrome c_1. In the Dcyc2 cyt1-CAPCH mutant, the holocytochrome c_1-deficient phenotype can be explained by the fact that both CCHL- and CC1HL dependent assembly pathways for apocytochrome c_1 are inactive. Whereas the absence of Cyc2p can account for the lack of cytochrome c_1 assembly via the CCHL pathway, we also envision that CC_1HL-dependent assembly of cytochrome c_1 is abolished in the 94cyc2 cyt1- CAPCH mutant due to the inability of CC_1HL to interact with CAPCH apocytochrome c1. This is supported by the fact that the level of assembled CAPCH holocytochrome c_1 is not influenced by the presence or the absence of CC_1HL, an indication that CAPCH holocytochrome c^sub 1^ assembly is dependent only on CCHL. Because the A-to-P modification in cytochrome c_1 was recovered in a search for mutations that suppress the absence of CC_1HL, it is conceivable that the P residue enhances the activity of CCHL toward apocytochrome c_1 while simultaneously preventing the CC_1HL from acting on its cognate substrate. The finding that the H, L, S, and T mutations and CC_1HL overexpression suppress the cytochrome c1 defect in a 94cyc2 cyt1-CAPCH context can be explained by a restoration of the CC_1HL dependent assembly of cytochrome c_1, presumably via enhanced interaction between CC_1HL and its cognate apocytochrome c_1 substrate.

The observation that the 94cyc2 cyt1-CAPCH mutant also displays a cytochrome c deficiency suggests that holocytochrome c assembly is also blocked as a result of an A-to-P mutation in the heme-binding site of cytochrome c_1. As apocytochrome c can be acted upon only by CCHL, we postulate that CCHL activity toward apocytochrome c is inactivated in the 94cyc2 cyt1-CAPCH strain. Two possible scenarios can be envisioned to account for this phenotype. In one model, CCHL can no longer act on apocytochrome c because CAPCH apocytochrome c_1 is trapped in a complex with the lyase. However, we view this model as unlikely on the basis of our observation that overexpression of CCHL in the 94cyc2 cyt1-CAPCH strain does not relieve the block in holocytochrome c assembly. The observation that the apoform of the molecule is not immunodetected in the 94cyc2 cyt1-CAPCH strain suggests that it has undergone proteolytic degradation, which does not appear consistent with a model in which apocytochrome c^sub 1^ is trapped in a stable CCHL complex. Instead, we favour an alternative scenario in which, in the absence of Cyc2p, CAPCH apocytochrome c_1 is engaged in a transient interaction with CCHL resulting in a deactivated enzyme that cannot assemble apocytochrome c. Note that the 94cyc2 94cc1hl strain also displays a dual deficiency in cytochromes c and c_1, similarly to the 94cyc2 cyt1-CAPCH mutant. Hence, we postulate that this deactivation is Cyc2p-dependent and takes places in situations where apocytochrome c1 maturation is dependent only on CCHL (i.e., when the apocytochrome c_1 heme-binding site carries the A-to-P mutation or in the absence of CC_1HL).

The proposal that the H, L, S, and T residues restore the CC_1HL dependent assembly of cytochrome c_1 further underscores the critical role of this residue in the maturation process. Interestingly, while the S residue at this position enhances CCHL-dependent assembly of cytochrome c_1, the presence of H, L, or T further decreases the intrinsic activity of CCHL toward apocytochrome c_1.

This suggests that the identity of the second intervening amino acid in the CXXCH motif is crucial in modulating the CCHL-dependent assembly of cytochrome c_1. The fact that CAPCH apocytochrome c_1 has become dependent on CCHL alone for its assembly indicates that the P residue is essential in determining the CCHL- vs. CC_1HL-dependent assembly of cytochrome c_1. It is likely that the nature of the residue is a determinant for enzyme-substrate interaction, but it is also possible that CCHL- or CC_1HL catalysed assembly requires a specific residue at this position for the enzymatic reaction.

Re-emerges in Dupont

Alternative corporate strategies of growth and of consolidation result in a changing focus for research and development. Over the past 70 years, the DuPont Company has passed through four recurring cycles, each lasting 15-20 years.

Within each cycle, the emphasis shifts from a period of discovery research to a later period of consolidation. It is the interplay between these phases that has shaped today's DuPont Company. The cause for this regularity in the cycles remains a puzzle. If history repeats itself, a new cycle is beginning and DuPont is on the verge of another burst of discovery research.

Observers have noted that basic technological innovations seem to occur in "bursts" or clusters in cycles of approximately 50 years, spurred by investor demand for new growth opportunities (1,2). Each "burst" is followed by a "long economic and investment wave," a period during which further discoveries and inventions are made extending and applying the new technology.

We have observed similar but shorter innovation patterns within DuPont. Since 1927, four cyclical periods, each between 15 and 20 years in length, have been identified. Within each, the emphasis shifts from a period of discovery, during which new products and related new businesses are created and expanded, to a period of consolidation, during which core manufacturing-based businesses are restructured to achieve efficiencies.

During the consolidation phase, scale-up of investments and commercialisation of new opportunities are focused but limited, and major cost controls are implemented in order to put new and current businesses on firm foundations.

It is the dramatic interplay between periods of growth, expansion and creation, and subsequent periods of consolidation and cost control, that has shaped today's DuPont Company. The bursts of productive discovery research have shown that good science can be translated into potential wealth, while the consolidation periods have demonstrated that this wealth can be captured and sustained.

The 17-Year Cycle

Many might wish that both growth and cost control can be done at the same time. Indeed, our experience indicates that these efforts are not always mutually exclusive. However, one of the two major corporate strategies appears to have predominated in each period. We can take our lessons from this history. A typical 17-year cycle might be characterised in the following way:

Years 1-3: There is dismay at a perceived shortage of good ideas. Studies are made of research to see if it can be made more productive. Various avenues for growth are explored. Employees in leadership positions have excellent skills in consolidation and cost control, but less experience in growing and developing new businesses.

Years 3-5: New ideas, including some previously held in abeyance, are brought forward. Discovery research flourishes, and new acquisitions in technology-based businesses are made. The groundwork for new opportunities is laid.

Years 5-7: Investments are made in the results of discovery research, and in new ventures and businesses. New venture managers learn to grow new businesses.

Years 7-9: Concerns are expressed that not all developments can be funded and that some new ventures have become a financial drain. Those now in leadership positions have not had significant experience in making difficult cost-cutting decisions.

Years 9-12: Funding is constrained for new developments. For R&D, the emphasis is on tailoring and improving current products and manufacturing processes. New-venture managers find other assignments or leave the company. Any new initiatives are conducted with very low profiles and funding is a significant challenge. Often outside funding is sought.

Years 12-15: The emphasis continues on cost reduction and reorganisation. Budgets are constrained, efforts focus on product quality, and some businesses are discontinued or sold.

Years 15-17: While costs are carefully controlled and more efficient methods of carrying out all business activity are found, there is talk of the need to grow and to increase revenues. However, these objectives are difficult to carry out.

We remain puzzled by the continued regularity of these cycles. Economic and market forces, world events and management changes may all play a role, but the 17-year period remains a mystery. We have attempted to determine if other companies have seen such cycles. Changes in R&D focus are common, but our search for examples of recurring cycles continues.

Commitment to Fundamental Research

At DuPont, the modern cycles began late in 1926, when concern over the lack of ideas for new products and businesses began to surface. A new commitment to fundamental research became evident in a budget letter of December 18, 1926, by the head of research, Charles Stine (5): We are including in the Central Chemical Departments budget for 1927 an item of $20,000 to cover what may be called, for want of a better name, pure science or fundamental research work.... The sort of work we refer to is work undertaken with the objective of establishing or discovering new scientific facts.... The volume of fundamental work is rapidly losing ground as compared with the volume of applied research. In other words, applied research is facing a shortage of its principal raw materials.

Stine proposed several new areas for researchers to explore, including catalysis, formation of colloids and polymerisation. He fought for his $20,000, received it and additional funds, and then spent the next few years protecting his program.

When the question of assigning scientists to fundamental research was under discussion six years later, in the midst of the Depression, DuPont's President Lammot duPont said, "It is more important to carry out research than to pay dividends".

There followed an innovative burst that lasted until the mid- 1 930s, laying the groundwork for nylon, neoprene, "Butacite" polyvinyl butyral (safety glass interlayer), moisture-proof cellophane, and even the chloride process for TiO2. This was a period of exciting discoveries in chemistry. Superb scientists, including Wallace Carothers, were recruited and produced astounding results. Acquisitions in chemicals, pigments, electrochemicals, and sporting goods further fuelled the company's growth. By 1935, DuPont's innovative drive fully exemplified its new slogan: "Better Things for Better Living Through Chemistry."

Three additional cycles of discovery bursts were to come, each providing significant new results for DuPont. Each period bore a remarkable similarity to the earlier ones.

1944 ushered in the second cycle, a period in which discoveries focused on polymer science. Academic science had, for the most part, dried up during the Second World War, creating a void in fundamental research. Research by industrial organisations was also restricted. The head of DuPont's corporate research argued presciently that "the country is about to enter a period of unparalleled scientific activity.... DuPont has to strengthen its research organisations and house them in modern research facilities. The Company's existing processes have to be improved and new processes and products developed; and fundamental research which will serve as a background for new advances in applied chemistry should be expanded".

The period between 1944 and 1952 saw the scientific groundwork laid for the subsequent commercialisation of "Dacron" polyester fibre (1950), "Mylar" polyester film (1952), urea herbicides (1953), and "Orlon" polyacrylonitrile fibre (1958). Studies of polymerisation of anhydrous formaldehyde were resumed in 1948 and prepared the way for the commercialisation of "Delrin" acetal resin in 1959. Early in 1953, the laboratory developed a fibre that was to become "Lycra" spandex fibre (1962).

The third cycle began in 1960. In that year, DuPont's executive committee expressed a concern over slowed growth and called for the research organisations to recommit themselves to the development of major new products. There followed a period of extensive discovery and diversification which has been called the "New Venture Era."

The 1960s saw the rapid commercialisation of over three dozen new products based on recent discovery research. These included "Teflon" FEP resin and film, "Tedlar" PVF film, "Vazo" polymerisation catalysts, "Uracil" and "Linuron" herbicides, "Elvax" vinyl resins, "Nordel" hydrocarbon elastomer, "Tyvek" spunbonded products, "Suryln" ionomer resins, "Kapton" polyimide film, and "Corfam" poromeric products. Each has its own story of the role played by discovery research.

Also in this period, in 1964, Stephanie Kwolek made the first liquid crystal polymer, and provided the basis for "Kevlar" aramid fibre, which was not to be fully commercialised for another 15 years. The R&D focus of the fourth cycle, which began in 1980, was on energy, life sciences, electronics, and imaging.

The company launched a major research audit in the early 1980s, aimed at identifying the key factors that stimulate technological innovation together with those that hinder its progress. Recommendations for improving the environment for innovation within DuPont were included. In 1981, a DuPont research director wrote that "the real problem lies in a substantially reduced backlog of innovative new technology to form the growth industries of the 80's and 90's, and in the increasing reluctance of industry, the government, and society in general to accept the reasoned risk-taking that led to our preeminent position in the world today". His words are even more relevant today.

During the period 198-1989, research in life sciences laid the foundation for the DuPont-Merck Pharmaceutical Company (1990). "Group Transfer Polymerisation" was developed in 1982 and exploited in a number of ways over the next 12 years. "Stainmaster" carpet was commercialised in 1986, and "Teflon" AF amorphous fluoropolymer was announced in 1989.

Periods of Consolidation

Between each of these bursts there occurred cyclical periods of consolidation, of renewed focus, of asset sales, and of extreme measures in cost control. During the second consolidation plan of each cycle, the focus was not on discovery research but on short-term support of the business, gaining cost savings from improved manufacturing processes and turning away from new opportunities.

Some of the major events of these consolidation phases include the following:

Cycle 1:

1940-1945: "Emphasis on the short-range research and development of products for wartime applications forces a significant reduction in the amount of fundamental research by the Chemical Department". 1942: Commitment to the Manhattan Project.

Cycle 2:

- 1952-1960: Polymer processing developed for "Lycra" spandex fibre, "Delrin" acetal resin.
- 1954: Chestnut Run Laboratory for product development and end-use research opens its doors.
- 1953: DuPont not interested in pursuing gas chromatographic instruments; Frank Martinez resigns to found F&M Scientific (now part of Hewlett-Packard).
- 1953: W L. Gore proposes program to solve fabrication problems with "Teflon:'
- 1958: W L. Gore resigns to exploit his invention of wire coatings. (W L. Gore and Associates is now a $1 billion business.)

Cycle 3:

- 1969: Divestiture of the dyes business.
- 1975: "The Company is supporting more new product and development programs than it can reasonably expect to commercialise"; departments then shift research emphasis toward short-term support and extension of existing businesses.

Cycle 4:

- 1989-1996: Divestiture or sale of Remington Arms, Med. Products, "Orlon," electronic connectors, explosives, methanol, polyethylene.
- 1989-1996: Major shift in emphasis from product to process research.

We have seen only one major counter-trend during the second half of the third cycle. George Levitt's first work on urea herbicides was in 1957; work was reactivated 17 years later (1974) for the development of the family of sulfonyl urea herbicides.

"In the 1970's, DuPont's agricultural chemicals business came under increasing scrutiny... Some members of the Executive Committee believed the managers of DuPont's Agchem business were taking too much money out of the business rather than putting it back into research.... Executives targeted research in agricultural chemicals for dramatic expansion".

In our most recent attention to cost reduction, 1989 to 1996, DuPont was reorganised with a focus on strategic business units. This design optimised business performance with a concentration on preferred markets for each business, together with rigorous control of expenses and capital. The R&D effort has been primarily focused on development. This approach has yielded the desired favourable impact on earnings and cash flow, but growth has at best only equalled that of the GNP.

Fifth Wave of Discovery Research

If history is to repeat itself in 1996-'97, DuPont is on the verge of a "fifth wave" of discovery research. We face a renewed challenge to bring forward major new ideas that can create new businesses and possibly new industries during the next 8 to 10 years. We can learn from history, but this does not mean that we can meet the challenge by relying on old historical patterns. Instead, we must develop new strategies that respond to a changed environment because the business environment today has changed in two critical ways.

First, there is strong continued pressure on individual businesses to maintain the emphasis on efficiency, and to develop products for well-known markets. Discovery research will probably not play a major role here.

Second, technical advances in communications have made possible a more rapid transfer of ideas, information and knowledge worldwide. Discovery research must be positioned not only to take advantage of new knowledge, but also to transform it to new practical opportunities and applications in advance of our competition.

This requires us to re-analyse the structure, focus and distribution of research efforts within and outside DuPont, and to develop new ways of accessing and linking with the larger body of scientific work beyond the company.

To capitalise on this period of opportunity, we have developed a strategic plan for discovery research which has six elements:

1. To provide direction and understanding for operating from a set of key beliefs that include the following:
   - Our competitive position and our growth potential are determined by the quality, depth, breadth, and exploitation of our technology.
   - We must stay connected to advances in science, mathematics and engineering; expansion of our technology depends upon these advances.
   - We have yet to capture the full benefits of our investments in science and technology. Opportunities abound close to home, as well as at a distance. For example, replacement of cellulosic fibres by non-cellulosics is just about complete, but replacement of metals by engineering plastics has just begun. Other more distant areas offer a host of opportunities for exploration, including new forms of carbon, biomimetic composites and combinatorial chemistry.
   - Our manufacturing processes are central to our competitiveness. Improved and new processes are required for growth, and to develop these, substantial improvements in process understanding, modelling and control must first be achieved.
   - Research processes themselves are changing rapidly and we must adopt many new practices: Information, computation, modelling, and analytical sciences are exploding. Complex problems can no longer be solved by a single person or discipline. Biotechnology is revolutionising the practice of chemistry, material science and agriculture. The research community is changing in focus and structure, and is now part of a global business.
2. To ignite discovery research by setting extremely far-reaching "unreachable" goals. A "futures" list might include immortal polymers, zero-waste processes, simultaneous monomer synthesis and polymerisation, 100 percent selective syntheses, anti-Markovnikov olefin amination, and "flexible" yet high-volume chemical plants. We should look for elastic coatings as hard as diamonds, elastomers as strong as steel, materials that repair themselves, chemical plants that are contained on one "chip," direct formation of garments from polymers, formation of the C-F bond without chlorine, and coatings that change colour on command.

3. To make corporate research the focal point for discovery research, but not the only point. Today, DuPont has a competitive advantage in its retention of a corporate research organisation, while other companies have greatly reduced their corporate research efforts or have moved them out to individual businesses. DuPont is reallocating resources to long-range research, increasing its emphasis on growth opportunities and on breakthrough opportunities that support current businesses. At the same time, support for maintaining current competencies and for renewal options for current businesses has been somewhat reduced in the corporate organisation.
4. To assure that corporate research efforts maintain a healthy tension between independence and connection. Research must be independent enough to provide freedom to create and to take chances, to push the research envelope, while at the same time it must remain cognizant of and connected to societal needs, and to the opportunities to expand current businesses and to build on global scientific advances.
5. To create a "growth network" of scientists across the company who can highlight and act upon new ideas as they come forward. This network can link specialised skills within DuPont with their counterparts in universities and other external resources. A focused scientific discovery effort inside the company must guide the formation of alliances and links with the frontiers of science, wherever they may be. This scientific network can be a valuable resource for evaluating any proposed acquisitions of new technology-intensive businesses to be undertaken in parallel with the discovery efforts.
6. To provide the new resources and research talent necessary to fuel the discovery research effort. Previous discovery bursts were made possible by such support. We expect that, for the fifth time in recent DuPont history, new ideas, discoveries and opportunities will materialise over the next 10 years. Some are already beginning to emerge-a new family of polyolefins and machines measuring disease-causing bacteria in food in hours rather than days or weeks (8,9); others still to appear will be completely unexpected.

Our challenge becomes one of transforming the products of basic scientific research into as-yet-unimagined products that can benefit society-"Better Things For Better Living Through Chemistry."

A Robust Strategy

Richard N. Cooper: I agree strongly with the main conclusion of this chapter by Klaus Lackner and Jeffrey Sachs, which is that there is no shortage of energy on the horizon. The chapter rightly emphasizes the substitution possibilities among different forms of energy, even at today's technology, and the abundance of total energy. The debate over the prospective exhaustion of liquid fuels takes many forms, but most are based on false premises, as the chapter usefully points out. The authors also argue that the world faces a serious problem in climate change, or, as they put it more generally, in the environmental constraints on energy use, and that it must be dealt with as quickly as possible. Furthermore, solutions are actually at hand with existing technology or are within reasonable sight, so there is some basis for dealing with the problem.

The chapter is an unusual one for a Brookings Papers panel. It is largely a primer on the role of energy in modern society, with a main emphasis on technology and technological possibilities. There is a fair amount of catalytic chemistry here, and some physics as well. Readers will have to brush up on their high school chemistry. The chapter performs a great service in being quantitative. This is, after all, essentially a quantitative topic. One cannot talk sensibly about energy alternatives without quantifying the possibilities. There are many attractive ideas out there, such as wind power, but when one looks quantitatively at the possibilities for mobilising them, it is clear that many can play only a niche role. The authors' focus on magnitudes is thus very useful in dispelling some myths or, more accurately, some wishful thinking. In this connection, however, I missed a more complete discussion of one potentially important technology, namely, the making of liquid fuel from biomass, both biomass grown for the purpose (and thus competing for land with food producers) and, more important, waste biomass such as corn stalks. The intellectual framework of the policy parts of the chapter involves stipulating some ceiling for atmospheric greenhouse gas concentrations, mainly $C[O_2]$. This threshold is not specified but is assumed to lie somewhere between two and three times the preindustrial concentration of about 280 parts per million. Thus the authors explicitly reject a cost-benefit approach to climate change, such as that developed by William Nordhaus, in effect assuming that the costs of climate change (and hence the benefits of mitigating it) become infinite beyond the ceiling. This assumption, I suspect, drives their call for early action.

The chapter constructs a baseline, "business-as-usual" scenario for greenhouse gas emissions over the next century on the basis of an assumed convergence of income per capita around the world on income per capita (or some fraction thereof) in the United States, which itself continues to grow: the gap is narrowed at a steady pace of 2 percent a year. This does not sound unreasonable until one realises that, along with the authors' other assumptions, it implies a growth in global income per capita of 3 percent a year until 2050. This compares with 2.1 percent a year during the half century 1950-2000. On historical experience, then, 3 percent is implausibly high. The implausibility is increased by using purchasing power parity (PPP) to calculate national and hence global GDP, so that the starting point is gross world product of $46 trillion in 2002, instead of the $31 trillion measured at market exchange rates. PPP, which, in effect, values output everywhere in the world at U.S. prices, gives much greater weight to agricultural output than does pricing at market exchange rates.

Since agriculture, which accounts for a large share of output in poor countries, typically grows more slowly than other sectors, giving it greater weight implies lower growth rates than the world is accustomed to. And calculated over a century, or even half a century, even small differences in growth rates can make a big difference. Applying the authors' assumed energy elasticity of 0.55 to a more reasonable annual growth in world income per capita of 1.8 percent (implying, with annual average population growth of 0.8 percent, an annual growth in gross world product of 2.6 percent, compared with the authors' 3.8 percent) would lead to $C[O_2]$-equivalent carbon emissions of 11.6 billion tons by 2050, compared with 17.0 billion in the authors' baseline, and 5.9 billion tons in 2002. Atmospheric concentrations of $C[O_2]$ by 2050 would then be under 500 ppm, rather than the 554 ppm in the authors' baseline; the figures for 2100 would also be significantly lower.

This adjustment does not, however, materially alter the authors' policy conclusions, given their focus on the need to avoid crossing a specified threshold; at best it provides a little more time. And, as the authors point out, in at least two respects their baseline projection is a conservative one. If coal liquefaction must take place earlier than they assume because of a more rapid depletion of conventional oil, or if automobile use in China, India, and other growing countries rises more rapidly than they assume, emissions will be higher.

The chapter places heavy emphasis, as does the current U.S. administration, on the sequestration of C[O_2], especially from power plants and other concentrated users of fossil fuels. Given the abundance of coal in the United States, China, and India, this is probably an appropriate emphasis for the next half century, and perhaps beyond. The authors suggest that the all-inclusive cost of carbon capture and storage (CCS) would be in the range of 1 to 3 cents per kilowatt-hour of electricity in an appropriately designed plant, raising the busbar cost of electricity (that is, the cost before distribution) by perhaps 50 percent. It is unclear where this and other cost estimates in the chapter come from; this one seems to be on the optimistic side. But the fact is that one cannot know how much it will cost until it is tried on a commercial scale, on which more below.

Costs are relevant, since, as the authors observe, nuclear power is an available alternative that does not emit greenhouse gases. Nuclear power has been economically unattractive in the United States during the past two decades, in part because of regulatory and legal delays and uncertainties. But with newer, standardised nuclear plants currently under design and a 50 percent or more increase in the cost of electricity from coal-fired plants, the economics of nuclear power could become much more attractive, especially in other countries but even in the United States.

Two problems of concern with nuclear power are how to store high-level nuclear waste (especially spent fuel rods) and the possible misdirection into nuclear weapons of the plutonium in spent fuel. I have never understood why so much effort has been directed at finding so-called permanent storage for nuclear waste; I would instead continue indefinitely the "temporary" storage used to date, which places the waste in secure, well-guarded concrete bunkers, where the containers can be watched and repaired if they corrode unexpectedly, and where the still considerable energy contained in them can be used if future generations discover how to do it economically and safely. The possible misuse of spent fuel for extraction of plutonium needs to be addressed through international agreement on a much tighter beginning-to-end nuclear fuel cycle than now exists. This will not be easy, but secure use of nuclear power depends on creating such a system, building on the current nonproliferation regime.

When it comes to CCS, as the authors note, it will be much more expensive to retrofit an existing plant than to design and build a new plant with CCS in mind. Once a large power plant is built, it will last for forty to sixty years. That argues for starting seriously, and soon, to include CCS in the design of all greenfield coal-fired power plants. The most active frontier now, when it comes to C[O_2] emissions, is China. China builds over a gigawatt a week in new power capacity, 70 percent of it fuelled by coal. China also has an aggressive program for nuclear power, an aggressive program for importing liquefied natural gas, and an aggressive program for hydroelectric power. The Three Gorges Dam, when the reservoir is completely filled in 2009, will generate 18 gigawatts of electric power. That is the equivalent of eighteen big nuclear plants or eighteen huge coal-fired plants.

But even with all that new capacity, China's demand for power is growing so rapidly that still more will be needed. The new coal-fired capacity that China is expected to install in the coming two decades exceeds what is projected for the United States and Europe together. In short, China is where the action is, and therefore that is where the focus on limiting climate change needs to be. China's new plants should be built with sequestration designed in. China will not agree to incur the extra costs; they will have to be incurred by the rich countries, especially the United States. And CCS is a promising but unproven technology. China's rapid construction program should be seen as a testbed for various ideas, to discover the most cost-effective way to sequester carbon, so that the technology can then be applied elsewhere as well.

The focus of this chapter is on climate change and, implicitly, on other environmental issues. But energy security is also an important issue. The world's growing dependence on the Persian Gulf for oil creates many misgivings. Thus, whereas the focus for climate change is coal, the focus for energy security is oil. The two policy objectives—mitigating climate change and ensuring energy security—run in parallel only a certain distance, and then they separate. Carbon sequestration, if the price can be brought down to 1 to 3 cents a kilowatt-hour, brings the two together, because coal liquefaction then becomes a viable substitute for oil. So that is another reason to develop sequestration.

The chapter observes that although the Kyoto Protocol is a "useful framework" for beginning to manage carbon emissions, it cannot do the job alone. The authors are too generous to the Kyoto Protocol. My own view is that it will divert attention from seriously addressing this issue for ten to fifteen years. The sooner we move beyond its framework, which cannot seriously reduce greenhouse gas emissions, the better.

My recommendation is a global carbon tax, as a charge for the negative externality of $C[O_2]$ emissions. Such a tax, which would be agreed upon internationally, but collected nationally, would encourage emissions-reducing actions across the board. I do not suggest that it would be easy to install, but the effort should get under way as soon as possible, and even if it does not succeed, the debate will bring global attention to the issue of climate change and shift the focus much more sharply toward those technologies that are likely to be productive.

William A. Pizer: Klaus Lackner and Jeffrey Sachs present a compelling case for worrying about future energy needs. They draw attention to an inevitable collision between the world's insatiable thirst for cheap energy and the increasingly threatening accumulation of carbon dioxide in the atmosphere. In turn, they use this collision to argue for two promising technological solutions over the next fifty years: $C[O_2]$ capture and storage from stationary sources coupled with gas-electric hybrid vehicles.

On a 100 year scale, they argue that the further use of electricity and hydrogen as energy carriers to replace fossil energy use at smaller sources (such as residential buildings and vehicles) will be the obvious technological solution. To move in this direction, they recommend that Europe, the United States, China, and India work together to ensure that future power plants meet a CCS emissions standard and that vehicle fleets meet more stringent efficiency standards. While noting the economic efficiency of tradable permits or taxes to achieve these ends, they suggest that a more practical alternative would be a performance standards approach. I tend to agree with most of Lackner and Sachs' main points. In particular, I believe a coming collision between global energy needs and concerns over global climate change are increasingly apparent. I also believe that CCS and hybrid vehicles are particularly promising technologies. And I believe that collaboration among Europe, the United States, China, and India is likely to be a more fruitful avenue for action than the existing United Nations framework and that technology policies (as distinguished from market-based approaches) could play a useful role.

Where I principally disagree is, first, with their intensive focus on CCS and hybrid technologies and, second, with their haste in dismissing market-based policies in favour of sector-based performance standards. My feeling is that the potential for nuclear power and renewable energy sources warrants a more balanced technological approach and, similarly, that technology policy should be viewed as a complement to a market-based approach, not a substitute for it. I also have a more subtle, almost philosophical disagreement with their overall framework, which tends to ignore the question of reasonable mitigation costs in favour of a laser-like focus on achieving zero net emissions by the end of this century.

Before addressing the paper's policy recommendations, however, I would note that the first third of the chapter provides an excellent synopsis of global energy demand, global energy supply, and global environmental concerns. The main points of this portion of the chapter are nicely summarised in the introduction: First, global economic growth will lead to increases in primary energy demand that are too large to be avoided through increases in energy efficiency. Second, there are no serious limits on fossil fuel supply over the next century, given the convertibility of coal into gas and liquid forms. Third, rising atmospheric concentrations of carbon dioxide from the combustion of fossil fuels pose a serious environmental threat—hence the collision of energy needs and environmental concerns.

Where I begin to diverge is with the authors' winnowing of promising technologies down to an almost exclusive focus on CCS and hybrids. While not ruling out other technologies they suggest, for example, a performance standard for power plants that could be met by nuclear or renewable fuels—they spend little time discussing these alternatives and the policies that might move them along. For example, they note that proven uranium reserves would be used up in a matter of decades if the global energy system went 100 percent nuclear. Yet a recent study by the Massachusetts Institute of Technology points out that increases in prices would likely produce significant increases in proven uranium resources, as occurs with virtually all mineral resources. One estimate suggests that a doubling of current uranium prices would produce enough supply to meet current global demand for fifty years. The same MIT report makes a host of practical suggestions, particularly surrounding waste and proliferation management, to expand the role for nuclear energy.

The chapter contains virtually no discussion of renewable energy sources, of which cellulosic ethanol and biodiesel are particularly promising. For example, a recent study by the National Commission on Energy Policy demonstrated that replacing half of oil demand with biofuels would have the same environmental impact as a doubling of fuel economy standards. Meanwhile the comparative political feasibility of renewable fuels versus increased fuel economy is evident in the energy bill enacted in 2005, which established a 7.5-billion-gallon renewable fuels standard starting in 2012 while ignoring fuel economy altogether.

Also, the costs of CCS and hybrids may not be as low as Lackner and Sachs suggest. In the case of CCS, they note that expenditures would amount to between 0.1 and 0.3 percent of gross world product, but they do not mention that the cost per kilowatt-hour amounts to a 15 to 45 percent increase in average U.S. electricity prices. In the case of hybrid vehicles, the authors estimate a positive net present value, relative to conventional vehicles, of $400 for a hybrid vehicle lasting sixteen years. But their analysis excludes the potential cost of replacing the battery ($2,000 to $3,000), uses a somewhat favourable discount rate and vehicle life (cars do last a long time but tend to be driven less as they age), and, perhaps most important, ignores the marginal net benefits of such a standard. Even if the average net benefits are positive, the marginal net benefits of doubling fuel economy are likely to be negative, perhaps significantly so.

None of this is meant to argue that CCS and hybrid technologies are not important, or even that they are not the leading technologies. Nor is it meant to suggest that such technologies should not be pursued because they may be more expensive than Lackner and Sachs indicate. However, they are just two among perhaps a half-dozen competing technologies, such as nuclear and renewable fuels, that may play an important role over the next several decades.

Ideally, policymakers would avoid having to choose among technologies by simply pricing $C[O_2]$ emissions, through a tax or tradable permit system, and letting the private sector work things out on its own. Various technologies could then compete without further government involvement. As the authors suggest, however, such policies are politically difficult in many countries.

In developing countries especially, it seems unlikely that governments would embrace such policies—indeed, if there was one trend evident at the recent UN meetings on climate change in Montreal, it was developing countries' uniform refusal even to discuss their accepting any form of emissions targets. Lackner and Sachs suggest sector-based performance standards as a logical alternative and next step. I would argue, a bit paradoxically, both that such standards at times do not go far enough to avoid political obstacles and that, at the same time, they are too quickly embraced by the authors as an alternative to market-based policies.

On the first point, one needs to recognise the significant obstacles to the performance standards approach in the United States. U.S. automakers oppose improved vehicle fuel economy standards both because of their own disadvantage, relative to their foreign competitors, in meeting them and because of their lack of expertise in hybrid technology, which is largely imported. Therefore a more practical policy—if one is motivated by that concern—is likely to require additional incentives to shift U.S. automobile manufacturing in a more fuel-efficient direction. In the area of electric power generation, a performance standard seems no more likely to be adopted than a tradable permit system, given the sector's experience and familiarity with tradable permits. Generally, the level of performance standards that Lackner and Sachs describe—a CCS standard for new power plants and a hybrid standard for new vehicles—seems far beyond what any nation is ready to embrace in the near future. Encouraging these technologies will require something more targeted and generous.

Beyond the United States, the enthusiasm of developing countries for a sector-based performance standard seems nearly as low as for an overall emissions target. A more likely scenario may be to develop a way in which climate-friendly policies can feed into market-based policies in industrialised countries. Such an approach was put forward at the Montreal meetings as a possible modification to the Clean Development Mechanism under the Kyoto Protocol.

This leads to my more important concern, which is that Lackner and Sachs fail to emphasize the need for some form of minimal market-based incentive, such as an emissions trading program with a somewhat weak target or price cap. Such an incentive will be useful both to get the right architecture in place as future policies may need to ramp up, and to provide some incentives in those areas where neither the sector- nor the technology-based approach applies. Recent research has emphasized the positive interaction of technology and emissions pricing policies.

All of these concerns relate to a somewhat philosophical question about how the problem should be framed. Lackner and Sachs establish as an imperative the development of a carbon-free future, and this in turn drives their emphasis on specific technologies that appear to be the more obvious solutions. Although they do not abandon a cost-benefit approach, they seem to loosely embrace the notion of a safe target for atmospheric C[O_2] concentration. They state that this approach is a reasonable one in "a situation where significant ecosystem changes due to anthropogenic climate change are assumed to have large but also unquantifiable consequences on global society." Such an approach pretends that, somehow, it is easier to pick a safe concentration level than to estimate climate impacts and their value. That does not ring true—although there are some threshold effects, many other effects are incremental, and even the threshold effects are notoriously difficult to pin down.

If we expend considerable effort to meet an atmospheric C[O_2] concentration target of 560 parts per million, but horrific effects occur at 550 ppm, what have we accomplished? The UN Framework Convention on Climate Change may have embraced the safe target approach for the sake of reaching an international accord, but it does not necessarily make sense (any more than the Clean Air Act's use of ambient air quality standards "to protect the public health" makes sense if dose-response functions are essentially linear). A particularly useful addition to the chapter would have been to include mention of recent summaries of estimated mitigation benefits, which provide a convenient mitigation cost benchmark for various technologies.

On the whole, Lackner and Sachs argue convincingly that action to address climate change is needed and that CCS and hybrid vehicles have an important near-term role. The weakness of the chapter is its failure to address other competing technologies with potentially significant roles. Their policy suggestion to focus on a smaller set of countries and not become overly hung up on market-based policy solutions is also useful; the weakness here is more a matter of detail. They fail to consider the significant value of having even a small market price on carbon and what the non-market-based policy alternatives might really need to look like. Finally, I find it disappointing that they provide no guidance on benefit estimates and instead emphasize safe concentration targets. To some extent this latter point has little consequence for near-term policies to encourage key technologies. In the broader debate over the design of tradable permit systems and the use of price-like mechanisms, however, such distinctions are important.

General discussion: Following Richard Cooper's suggestion of a global carbon tax, other panelists discussed the need for policies that would lead to achieving the paper's goals. Benjamin Friedman noted that the paper had outlined a number of desirable changes to present practice and technologies, but that its proposals also carried some of the flavour of a command economy. It was not clear, he argued, what specific policies would provide incentives for private endeavour to lead in the directions the paper favoured. Alan Blinder agreed but added that there may be no good policy answer for some of the needed changes. No global government exists to impose a global carbon tax, and overcoming the externality appears to be a nearly insuperable problem.

Gregory Mankiw agreed on the need for a global carbon tax and argued that the cap-and-trade system envisioned by the Kyoto Protocol was equivalent, from an efficiency standpoint, to such a tax. However, the cap-and-trade system would inevitably involve cross-national transfers, which would surely create political difficulties, whereas a carbon tax could be imposed by each government individually, and no transfers would be necessary.

David Laibson disagreed with Mankiw's suggestion, arguing that the developing countries, especially, would not go along. He agreed with Friedman that the problem was how to design specific, incentive-based mechanisms to achieve the desired goals. What the paper should address is which of various alternative mechanisms would provide incentives for the governments of each of the roughly 200 countries in the world to adopt a carbon tax. Such a mechanism would entail transfers from the developed world, which has the greatest interest in the issue, to the developing countries, which can least afford to do anything about it.

To Friedman, the authors seemed quite optimistic about solar energy for the long run, whereas in popular discussion it does not figure as prominently as either fossil or nuclear energy. He asked whether the associated technology is improving and the cost decreasing. Klaus Lackner replied that solar power could potentially provide all the world's energy needs but is very expensive. Mass production could significantly lower the cost, but, beyond that, improved technology would be needed if solar power is to greatly expand its present role.

Blinder remarked on the authors' observation that hybrid automobiles are cost efficient over a fifteen-year life span. Most people replace their cars every three or four years and (although with perfect capital markets this should not matter) tend to behave as if they face extremely high interest rates. This, along with the high cost of replacing the batteries at intervals, Blinder concluded, poses a significant obstacle to the spread of hybrid automobile technology.

Robert Gordon professed surprise on hearing Cooper refer, in his discussion of the paper, to economies of scale in coal-fired power plants as if these were somehow unlimited. Gordon's own recently published study of electric power generating plants had concluded that the industry had reached a plateau of efficiency and productivity in the late 1960s, due both to the constraint of supercritical pressure levels, which act like the sound barrier to limit scale and efficiency, and to the physical constraints imposed by federally mandated scrubbers and other antipollution equipment. Gordon doubted that subsequent technological improvements could have overcome these fundamental barriers.

Chapter 6

Multiple Local Sequence Alignment and Gibbs Sampling Strategies

The linear biopolymers-DNA, RNA, and proteins-are the three central molecular building blocks of life. DNA is an information storage molecule. RNA has a wide variety of roles, including a small but important set of functions. Because RNA can play so many roles, it is believed to be the molecule from which life began. Proteins are the action molecules of life, responsible for nearly all of the functions of all living beings and forming many of life's structures. The diversity of functions performed by proteins is extraordinarily broad. Although the methods we describe here have been successfully applied to all the three biopolymers, to date they have been most widely used for the characterisation of proteins, the focus of the application described here. A protein is an unbranched, linear, heterogeneous polymer composed of a sequence of 20 types of amino acids. An amino acid is composed of a peptide and a side chain residue.

The peptide is identical in all but one of the amino acids, proline; however, all 20 side chain residues are unique. The peptides link together to form the peptide backbone chain, with the side chain residues attached at regular intervals. Thus a protein may be specified in a one-dimensional representation by giving the sequence of its residues. Most proteins fold to a unique densely packed three-dimensional shape. Because of the dense packing, a figure that includes all the atoms in a protein is extremely cluttered.

Probably the most important and impressive job performed by proteins is the catalysis of biochemical reactions. Enzymes, which are complexes of one or more proteins, are extremely efficient catalysts, with efficiencies several orders of magnitude greater than the best chemical catalysts. The fact that chains of 20 simple molecules can carry out such a broad spectrum of catalytic function so efficiently makes them arguably the most impressive system in all of the natural world. In large part enzymes owe their impressive catalytic efficiency to the exacting three-dimensional structures to which they spontaneously fold.

Enzymes often recruit other molecules, called cofactors, to assist in their catalytic action and perform their catalytic function through the precise interaction of the enzyme and perhaps a cofactor with the molecule to be chemically transformed, its substrate. This interaction requires the enzyme to bind both the substrate and the cofactor.

Many cofactors are derivatives of vitamins and typically contain several chemical groups. In the example used here, three cofactors are involved: nicotinamide adenine dinucleotide (NAD), nicotinamide adenine dinucleotide phosphate (NADP), and flavin adenine dinucleotide FAD), which are derived from niacin, niacin, and riboflavin. All three cofactors contain two phosphate groups, composed of a phosphorous atom and four oxygen atoms. In the example we focus on the binding of the enzyme to the phosphate groups in these cofactors.

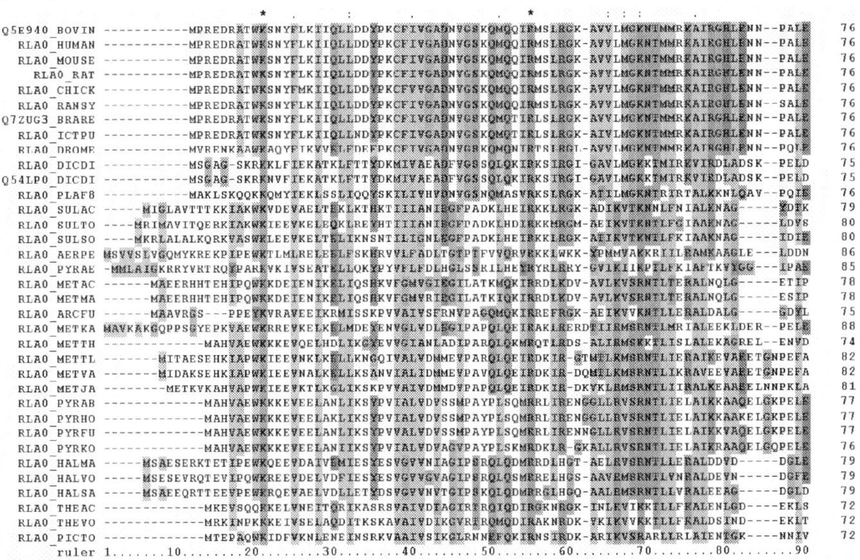

Figure: *First 90 positions of a protein multiple sequence alignment of instances of the acidic ribosomal protein P0 (L10E) from several organisms. Generated with ClustalX.*

The critical interactions between an enzyme and a substrate, cofactor, or other bond molecule involve only a limited subset of its residues. These critical residues occur in a few segments of the protein chain. In the example, Applications & Case Studies a few segments are involved in cofactor binding, but only one such segment in each binding site is involved in binding to the important phosphate groups common to all three cofactors. The sequence of residues in the binding segments of these proteins is constrained by the requirement that the residues fold to this distinctive shape and interact properly with the phosphate groups. Surprisingly, however, there is considerable variability in the sequence of residues that can meet these constraints.

Variations and Commonalities Among Functionally

Related Proteins

Variability in the biopolymer sequences arises from mutations that occur during evolution. In most cases, today's sequences are believed to be the progeny of common ancestral sequences and thus are not independent. The mutations happen at a relatively slow rate, so species that are not too far apart in evolutionary time (e.g., man and monkey) are strongly correlated. Homologous proteins within a single organism arise from a duplication of a common ancestral sequence and are passed on in parallel. When biopolymers have been subjected to a limited amount of evolutionary change, their commonality stems primarily from their mutational history. Such closely related sequences are relatively easy to align. The focus here is on the more difficult case that arises when the sequences have been subjected to extensive change and common patterns are subtle.

The other force at work during evolution is natural selection. Intuitively, natural selection limits which mutations survive to be passed on and, therefore, defines the common model for structurally and functionally related proteins. If a mutation is deleterious, then the mutant protein is eliminated from the population. The force of selection on a protein thus arises from its ability to function properly, and function depends on structure. Among distantly related sequences, common features stem primarily from the common set of structural or functional constraints rather than from evolutionary history. Thus in the limit, the sequences may be treated as independent observations. Statistically, the model described later and some other statistical methods that have successfully aligned large numbers of sequences assume that sequences under analysis are independent given the common model. In many cases this assumption can be very closely achieved through careful selection of the data set. In practice we have found that the methods we propose here often work well even with substantial departures from this assumption.

There are four classes of mutations: point mutations, insertions/ deletions, transpositions, and duplications. A point mutation occurs when a residue at a given point in a sequence is mutated to a new residue type. In our example we focus on proteins that bind either NAD, NADP, or FAD as a cofactor. To achieve binding there must be a force based on the energetic requirement of the protein to hold the cofactor in place. This requirement imposes constraints on point mutations in the binding site. The relationship between energetic constraints and frequencies forms the basis of statistical mechanics, pioneered by Gibbs and Boltzmann. There is an analogous relationship for residue frequencies subject to random point mutations, which forms the foundation of the models used here. From a statistical modelling perspective, this relationship is extremely valuable, because it permits us to translate from the language of physics and chemistry that governs molecular bahaviour to the language of statistics.

Multinomial models have been used successfully to capture the variability and the limitations imposed on these sequences. Because the most important interactions of residues in a binding site area with the cofactor, substrate, or the environment rather than with one another, motif models that assume independence of the positions provide a good first approximation. Thus we use product multinomial models to describe the binding sites. The remainder of the enzyme forms a scaffold that supports the substrate binding site. Because in distantly related proteins these scaffolds can vary greatly, there is often little in common in the remainder of the protein of interest to us. Consequently, they can usually be described by a simple iid or low-order Markov sequence model.

The other three classes of mutations - insertions/deletions, transpositions, and duplications - result in changes in the length of the sequence or in reordering of the sequence. Let us consider in more detail the effect of deletions with specific attention on the segments involved in cofactor binding. A deletion mutation will remove a segment of a protein sequence, and the resulting two adjacent fragments of the protein will be shifted to form a continuous chain. If the deletion is "upstream" of the cofactor binding segment, then this segment will be shifted to the left and thus misaligned with respect to its predecessor.

Insertions operate in an analogous manner but add sequence segments. Transpositions move a segment to a new location in the sequence. Duplications replicate segments and then insert them in new locations. On the other hand, the stringent geometric requirements on the binding segments can rarely tolerate internal length changes, and thus in most cases can be modelled well as segments with no internal misalignment.

Missing-data Methodology for Alignment Problem

The variations of the locations of the phosphate-binding segments provide a good illustrative example of the effects of these mutational processes. The underlined segments, each of which corresponds to a substrate binding segment, occur at various different locations in each sequence. Unfortunately, direct data that specify the locations (i.e., the indices of these binding segments in each sequence) are not available. The data available for analysis are without underlines or any other indication of the positions of the common segments in these sequences. This type of misalignment of sequences is the distinguishing characteristic in the analysis of biopolymer sequence data.

Lawrence and Reilly (1990) recognised that it is constructive to consider alignment variables as missing data and presented an expectation maximisation (EM) algorithm for the alignment of gene regulatory sites in DNA sequences. Cardon and Stormo (1991) extended this approach to allow for one small gap in the middle of each segment. Liu (1994) showed that an iterative sampling approach could be used to advantage in the identification of gene regulatory sites, and described a Gibbs sampling algorithm for its solution. The collapsing theorem developed there has been found very useful for these problems. Lawrence et al. (1993) have described a Gibbs sampling algorithm for local multiple alignment of protein sequences. Alignment methods based on hidden Markov models (HMM) have recently been described. Other statistical approaches for sequence alignment have also been described.

Multinomial Models

Given a set of sequences, $[R_1]$, $[R_2]$,...., $[R_K]$, without underlining, our goals are twofold: (1) to identify the substrate binding segments, which we call "elements," and (2) to estimate the parameters of the product multinomial model that describes the collection of aligned elements. Here we call this common model a "motif." We call position a "site" if it is described by the motif and "nonsite" otherwise.

Three classes of mutations result in misalignment of common elements. Another data structure, constituting the alignment, is a set of positions [a_k] for k from I to K, for the starting positions of the elements within each sequence. Regarding the a's as missing observations, we can effectively apply missing-data methodology. One of our goals will be to identify the "best," defined as the most probable, motif and alignment. To help fix the idea for the material that follows, consider the following coin-tossing analogy. The game is played with L coins, say 50. (L - J) of these coins, say 40 (plain coins), all have the same probability of heads, not necessarily a half. The remaining J = 10 special coins have probabilities of heads different from the plain coins and different from one another. On each of K independent trials, corresponding to K observed sequences in alignment problems, the coins are shaken in a tumbler and laid out in a row. It is known that the 10 special coins are positioned in a contiguous block of 10, with the special coins always in the same order is each trial. But their actual positions vary from trial to trial and are not observed. The player is challenged to estimate the probabilities of the 11 types of coins and to specify the locations of the special coins in each trial. This game corresponds to an alignment problem with p = 2 (two-letter alphabet), J = 10, and [L_1] = ... = [L_K] = 50. The associated parameter [theta]'s describe the unknown probabilities of showing heads for the J + 1 types of coins.

Product Multinomial Model With Dirichlet Priors

For a set of categorical data R = {[r_1]..., [r_m], where each ri takes values from a p-letter alphabet, we define a counting function h such that h(R) = ([m_1], [m_p])[T], where [m_k] is the total number of kth type letter observed in R. For protein sequences, the h function counts the numbers of different types of amino acids in set R and results in a 20-dimensional column vector. It is noted that the function h has a nice additive property. If another data set R[minutes] = (r[-1],...., r[minutes]$_n$) is given, for example, then h(R) + h(R[minutes]) = h(R [direct sum] R[minutes]), where the left side is just the ordinary addition for vectors and R [direct sum] R[minutes] indicates combining the two categorical data sets R and R[minutes].

Alignment When the Number of Elements Is Given

The use of iterative sampling for Bayesian missing-data problems was first undertaken by Tanner and Wong (1987) and Li (1988). This sampling approach and its extensions have recently become a topic of great interest in Bayesian statistics. Gelfand and Smith (1990) illustrated its connection with a more general method - the Gibbs sampler - and greatly popularised the idea. The formula implies that conditional on the fixed sites of the elements in the rest of the sequences, the probability that the element in sequence k starts at position i is proportional to the likelihood ratio of its being a site to its being a nonsite. We tested both the exact formula and the approximation in our algorithm. We found no observable discrepancies between the results obtained from using the two different formulas.

Generalisations to Multiple Elements and Multiple Motifs

The foregoing method can be generalised to the case where there is more than one copy of the common element in each sequence. For example, suppose that there is only one sequence (i.e., K = 1) but there are m copies of the common element to be aligned. Let A = $\{[a_1,1],...,[a_1,m]\}$ be the set of starting positions of the m elements.

Fragmentation Model

Previously, we modelled the motif as a contiguous block of width J residues. But usually not all the positions within the contiguous block of an element are important for protein structure and function. To accommodate this feature, we may wish to select J < W positions in an aligned block of width W residues in each of the K sequences to form the motif model. This model automatically takes care of the phase-shift problem noted by Liu (1994). Let [C] = $([[\Delta]_1],...,[[\Delta]_w])$, where $[[\Delta]_w]$ = 0 or 1, and = J. The [Δ]'s indicate which positions (or columns) within the aligned block are included in the motif model. The vector [Δ] is in fact another high-dimensional parameter and can be treated as missing data as well. Treating [Δ] as missing data implicitly assumes a flat prior on all its possible values; that is, a priori [Δ] takes any 0-1 vector satisfying the constraints with equal probability $1/([^W{}_J])$.

To sample from this distribution, we proceed in two steps. First, we sample to find a contiguous block of J residues. This is equivalent to assuming that $[[\Delta]_1]$ = ... = $[[\Delta]_J]$ + 1 and all the rest are zeros. Second, we use a Gibbs sampling strategy for identifying the best positions. The way to achieve this is to permit the exchange of one of the J positions in the motif model for one of the (W - J) nonsite positions within the aligned block. We draw one of the J positions to take out of the model completely at random, followed by the sampling of a replacement position from one of the (W - J + 1) unoccupied positions where sampling is now in proportion to the ratio of the site to nonsite gammas in Equation. It is typical in our applications that each position column has only tens of residues, whereas all the nonsite residues add to several thousands.

Weighted Prior For Fragmentation

In previous subsection we modelled a priori that $[\Delta]$ takes any W-long 0-1 vector satisfying $[[\Delta]_1] + \ldots + [[\Delta]_w] = J$ with equal probability $1/([^W{}_J])$. But real protein sequences favour short motif and thus suggest that the "span" of the element should be small. Let $[w_0] = \min\{w: [[\Delta]_w] = 1\}$, and let $[w_1] = \max\{w: [[\Delta]_w] = 1\}$. The span of an element is defined as $[w_1] - [w_0] + 1$ whose smallest possible value is J. The rationale is that given the length of the span, there are ways of assigning zeros and 1s for positions within $[w_0]$ and $[w_1]$. In other words, this prior assigns equal probability on all possible $([w_0], [w_1])$, the span of the element. We have tried both prior distributions in many test examples and found that the latter is significantly better than the former in the sense that the sampler identifies the correct locations of the elements more often. This weighted prior for fragmentation is used in the dinucleotide-binding example.

The Rank Test of Significance

The assessment of significance for a sequence alignment is a known difficult problem. It is particularly difficult when the residue frequency parameters must be estimated from the data. The problem is perhaps easily addressed for standard problems where a full Bayesian analysis can provide not only a point estimate, but also the whole posterior distribution. But in our problems, the sample space of the joint alignment A is discrete with an astronomical size. In the example, the size of the space is $[2^{36,000}]$. The resulting posterior distribution of A is typically spiky with many local modes, even when the sequence data are randomly generated. When likelihood methods are used, the likelihood ratio test is useful only when sample sizes are larger than that frequently available for protein alignment problems. Comparison of results with multiple (typically 100) random shuffles of the data is frequently the only available tool for the assessment of significance. Here we describe a ranks test that compares the sequence to a single control data set. The only restriction on the selection of the control data set is that the lengths of the sequences used in the control be equal to those in the test data set. Next we describe the procedure for Bernoulli sampling. For multinomial sampling, the procedure is the same, except that a control sequence is appended to each test sequence.

The rationale behind the test is that under the null, the elements are equally likely to be solicited from either the study or the control sequences. The circumstance of our test is not the same as that of the classical Wilcoxon test, where typically two paired samples are involved. But conditioned on the total number of elements found, the same exchangeability argument as that of Wilcoxon's can be applied. In the case that no elements are found in the control data set, for example, the significance will be simply. Because the Bernoulli sampler excludes any overlap of two elements, our test is actually slightly conservative. (The sampler will be pushed towards the control data if the elements are too crowded in the test data.)

The generation of negative control data sets is a well-recognised issue in the field of computational biology. Lipman, Wilbur, Smith, and Waterman (1984) showed that some local dependence of a sequence can affect the statistical significance of a similarity. Thus one may want to use negative control data sets other than the randomly shuffled set to guard against possible artifacts. Our test method is not restricted to random shuffled sequences and can be carried through for any carefully chosen negative control data sets that accurately reflect one's null hypothesis. In this sense, our method is also open to possible abuses. Another issue concerns the implication of our reference p value, which would perhaps have been different had another control data set been generated. Note that this kind of fluctuation is the characteristic of all Monte Carlo tests. Following Hope (1968), our reference p value based on one shuffled control set is a valid p value. That is, if we calculate that the reference p value is [α], then the probability of making a type I error by rejecting the null hypothesis at level [α] is equal to [α]. Hope (1968) also pointed out that generating more reference sets can increase the power of the test.

Example: Dinucleotide-Binding Proteins

NAD, NADP, and FAD are cofactors used by several different enzymes. Neuwald and Green (1994) described a set of 91 distantly related proteins that are components of enzymes that bind one of these cofactors. This set contains no pair of sequences to be found significantly related using the BLAST algorithm. We replaced two of the proteins in this collection with two other dinucleotide-binding proteins whose three-dimensional structure is known. We made this replacement in a manner that assured there were no significantly related pairs. Several of the proteins in this set are components of enzymes complexes of several proteins. Consequently, not all of these 91 proteins are expected to directly bind one of the cofactors. Furthermore, some of these sequences may contain multiple dinucleotide binding sites.

The prior Dirichlet distributions for the frequencies of the 20 residue types for all site and nonsite positions were chosen so that the expected values of the frequencies were proportional to the composition of the entire data set. The pseudocounts (i.e., $[[\alpha]_1] + \ldots + [[\alpha]_p]$ or $[[\beta]_1 j] + \ldots + [[\beta]_p j]$) were chosen as 9, 10% of the total number of sequences. Because we expected some proportion of the sequences to contain an element, we chose the parameters of the prior beta distribution in such a way that the expected number of elements was 45, corresponding to half an element per sequence, and with a standard deviation of our default value of 6, 2/3% of the mean.

This gave us a = 180 and b = 144,984. In this example the results do not change remarkably for different choices of (a, b) ranging from expected 1/3 element per sequence to one element per sequence. Because in this example the width of the motif is unknown, we applied fragmentation using the weighted prior with the defaults of J = 13 and W = 130. We applied the signed ranks test with the same prior distributions as those described earlier, except that we modified the beta parameters to correspond to a doubling of the expected number of elements. The sampler applied to the data set including the controls reported 38 elements that were sampled 50% of the iterations after convergence. Seven of these came from the control data set with a total rank of 55, which corresponds to a Z value of -4.6 using normal approximation with correction. In all cases we judged convergence in accordance with the procedures described by Lawrence et al. (1993).

Aligned elements from these 91 sequences were identified by the sampler. Notice that the fragmentation model has selected a motif that spans 26 residues, and that this motif corresponds quite well with the strand-helix-strand motif of ADHE. It also corresponds to the locations of the known dinucleotide-binding strand-helix-strand structure in the two other proteins in this set for which such structural data are available. Furthermore, this motif corresponds nearly perfectly to a sequence fingerprint model for dinucleotide-binding segments derived by superimposing the structures of several of these proteins.

In protein motifs, not all of the positions are equally important. All three positions are most frequently glycines, the smallest and most flexible of the 20 amino acids. In this motif, these three positions, play important distinct roles in the binding of NAD. The flexibility of a glycine residue at position 199 of ADHE (motif position 5) permits the protein chain to make the sharp bend. The small size of a glycine at position 201 (motif position 7) permits the motif to closely approach the phosphate groups of NAD to which it binds.

The small size of the glycine at position 204 (motif position 10) permits the first strand of the strand-helix-strand structure to closely approach the helix. Position 3 of the motif (position 197 in ADHE), is the fourth most distant position. One of three residues (valine, isoleucine, or leucine) is required at this position. These three residue types are often found in the interior of a protein and help to stabilise a protein's structure. In this case this position helps to hold the first strand of this strand-helix-strand motif in position against the helix. In most cases sequences available for analysis do not arise from a process that mimics independent samples from a common model, but rather they emerge by evolving from common ancestors.

Some methods for sequence alignment that incorporate evolutionary history have been described, but the results from these in multiple sequence alignment have so far been limited. When sufficient diverse data are available, as in our dinucleotide-binding protein example, the independence assumption can be very closely achieved by removing highly correlated observations from the data set. In cases where there are insufficient data to take this approach, a method that can simultaneously align a large set of sequences (especially sequences that include subtle relationships) and account for the correlations stemming from evolutionary history awaits future development. Although theory addressing these correlations is important, in practice we have found that the methods we propose here often work well even with substantial departures from the independence assumption. To date, our experience and that of others using these methods (S. Henikoff, personal communication) indicate that departures from the independence assumptions have nearly no impact on the methods. Some basic similarities of this work with previously reported methods should be noted. Existing Gibbs sampling and EM algorithms for multiple sequence alignment share the treatment of alignment data as missing. Two EM approaches have been developed: one that finds alignments of ungapped elements, the other in the form of hidden Markov models (HMM) that permit gaps between any two neighbouring residues in a sequence. Because the joint distribution of multiple elements in each sequence must be explicitly estimated, the first approach has been limited to problems with only one or two motifs. The sampling methods permit us to escape this restriction by exploring the joint distribution through sampling the complete set of conditionals.

When sequence evolution has not been subject to duplications or transpositions, the order of the residues in the sequences will not be altered, and the sequences are said to be collinear. This collinearity induces a Markov relationship on the alignment, which forms the basis of a popular dynamic programming algorithm for the alignment of a pair of sequences. HMM's exploit this same characteristic to explore the joint alignment distribution, which allows gaps anywhere in the sequence. While increasing their modelling flexibility, the unrestricted HMM's have to pay the price of a loss in specificity and an increase in computational complexity. Of course, their use of collinearity restricts them to problems where the orders of the motif elements have not been transposed. The sampling methods discussed here do not have this restriction. All currently available HMM algorithms are based on the EM approach, which, because of its deterministic nature, can become trapped in a local mode. The ideas presented in this chapter can be useful for building full Bayesian models and Gibbs sampling strategies for the HMM's.

Choosing the pattern width J in our analysis corresponds to the well-known problem of model selection when there are changes in dimensionality of the freely adjustable parameters. The celebrated Akaike information criterion (AIC) and Bayesian information criterion (BIC) have been useful in many applications. Unfortunately, these criteria did not perform well in selecting those pattern widths that identified correct alignments in data sets with experimentally determined "known solutions." When fragmentation is not used, the "information-per-parameter" criterion proposed by Lawrence et al. (1993) still seems to perform well, although this criterion biases toward elements with strong ends. When fragmentation is used, our experiences suggest that a single default setting of $J = 13$ and $W = 130$ works well for most protein alignment problems. Clearly, this single default is able to identify most motifs because (a) 13 columns are sufficient to identify most protein motifs, (b) the weighting scheme in Section 3.5 effectively eliminates the need to constrain W, and (c) the single default focuses the sampler's attention on the J positions that are farthest in relative entropy from the background frequencies. Consequently, only those positions in a motif with relatively small entropy distance from the background frequencies are excluded from the model. In some circumstances a biologist will have prior information to choose a value for J that differs from the default.

There are several biopolymer sequence data bases. Records in these data bases begin with the entry of the biopolymer sequence. Additional information about the biopolymer is sometimes added to the record as annotations.

Among other things, these annotations can conclude experimental evidence that a protein binds a specific ligand and evidence of the location of the binding site in the sequence, which usually requires a known three-dimensional structure. Because these data bases are now in a state of rapid growth, the state of characterisation for many sequences is very limited. Thus, as is typical, it is not possible to verify the binding of a cofactor for most of the 91 sequences in our example, much less to identify the specific sites of binding. In this example the best evidence for convergence to the "correct" alignment is the excellent agreement of our motif with the fingerprint model reported by Wierenga et al.

There are a few approximation formulas throughout Sections 3 and 4. They are not necessary, and the corresponding exact formulas are provided. These approximations speed computation somewhat, but more important, they provide statistical insights into the meanings of iterative sampling procedures.

The models discussed in this chapter can be viewed as mixture models, and the problem addressed by the Bernoulli sampler can be regarded as a classification problem. More precisely, we can treat each position in every biological sequence as sampling from a mixture of "element" and "nonelement." Our experience demonstrates that this simple viewpoint is especially constructive. Gibbs sampling methods have been shown to be useful for these models and have been particularly constructive for the identification of subtle relationships among protein sequences. Here we provide the rigorous Bayesian foundations for the application of these sampling methods for multiple sequence alignment. In addition, we show how to extend these methods by relaxing the specification of the number and size of motifs. Furthermore, we develop ranks test to judge the statistical significance of a multiple alignment.

Organisation of Matter

Fredrick Engels formulated some theses of a dialectical philosophy of nature that remain very topical today:

"The real unity of the world consists in its materiality" (1987a, 41).

"The basic forms of all being are space and time, and being out of time is just as gross an absurdity as being out of space" (1987a, 48-49).

"Motion is the mode of existence of matter.... Matter without motion is just as inconceivable as motion without matter. Motion is therefore as uncreatable and indestructible as matter itself.... Motion therefore cannot be created; it can only be transferred" (1987a, 55-56).

The human mind is the highest product of organic matter (1987b, 335; 1990, 369).

"Nature does not just exist, but comes into being and passes away" (1987b, 324); it "has its existence in eternal coming into being and passing away, in ceaseless flux, in unresting motion and change".

Matter is "eternally changing, eternally moving,... we have the certainty that matter remains eternally the same in all its transformations, that none of its attributes can ever be lost, and therefore, also, that with the same iron necessity that it will exterminate on the earth its highest creation, the thinking mind, it must somewhere else and at another time again produce it" (1987b, 335).

Nature forms a system, an interconnected totality of bodies that react on one another; this mutual reaction constitutes motion.

"The basic form of all motion is approximation and separation, contraction and expansion—in short, the old polar opposites of attraction and repulsion" (1987, 364). This can today be interpreted in such a way that all forms of matter are in continual motion; they produce chaos, and they also produce order from chaos, and hence higher levels of organisation. The dialectic of attraction and repulsion is a description of dynamic movement that produces emergent qualities on higher levels of organisation.

"Matter is nothing but the totality of material things from which this concept is abstracted.... Words like matter and motion are nothing but abbreviations, in which we comprehend many different, sensuously perceptible things according to their common properties" (1987b, 515). Matter is an abstraction in the sense that we abstract from the qualitative differences of things and combine them as physically existing in the concept of matter (533-34).

Matter is the totality of objects that constitute reality and is itself constituted in space and time by an interconnected totality of bodies that react on one another (motion)—that is, they repulse and attract each other. Motion is the mode of existence of matter in space-time. Matter is an eternal process of becoming and passing away, a ceaseless flux; it is uncreatable and indestructible. Matter is the totality of objective, really existing systems that are interconnected and subject to different physical laws.

Matter develops dialectically, and this development produces various forms of matter that have emergent qualities that distinguish these different forms. Matter can exist independently of human consciousness. Consciousness is not a necessary result of the development of matter, but it has historically emerged from it. As an activity of the thinking brain and as part of the human being, it thus forms part of a specific organisational level of matter that we can term the level of human beings. The material unity of the world means that the motion of matter results in a natural hierarchy of relatively autonomous forms of movement of matter, where each level has new, emergent qualities that cannot be reduced to lower levels or an assumed primary form. Time is an expression of the irreversible changing state of matter. Movement in time means movement in space and vice versa. Both space and time express the permanence of change that is a fundamental property of matter. Matter permanently organises itself and produces an irreversible sequence of states.

Attraction and repulsion are the essence of matter; as polar opposites they are "determined by the mutual action of the two opposite poles on one another,... the separation and opposition of these poles exists only within their unity and inter-connection, and, conversely,... their inter-connection exists only in their separation and their unity only in their opposition".

Energy is the measure of the capacity of a physical system to undergo change; it is an attribute of matter. Energy is not something external to matter, but is inherent in matter. Physical conceptions, such as Heisenberg's conception of the field as the source of particles, the assumption of quarks as elementary particles, etc., show that the source of existing forms of matter is itself material and that the unity of the world is its materiality.

In contrast to dialectical materialism, mechanical materialism has been invalidated by modern physics. Dialectical materialism's assertion that the world is in constant flux and process is continually borne out. The basic hypotheses of Marx and Engels about the dialectics of matter still remain topical. Complementarity does not mean a dualistic, but a dialectical, relationship of wave and particle.

Energy and information do not exist outside of, nor are they external to, matter; they are specific aspects of the movement and development of matter and as such are integral aspects of the world. The Middle Ages were dominated by a religious conception that considered matter as a creation of God. This was questioned by pantheistic conceptions such as that of Giordano Bruno, who considered God as an eternal force immanent in nature. The Newtonian worldview was characterised by its belief in the absolute immutability of nature and a reductionist methodology. Nature was considered as a conservative system that remains stable from its beginning until its end. Organic matter was reduced to mechanics. French materialism of the eighteenth century as well as the "mechanical" materialists of the nineteenth century were influenced by this worldview. The human being was considered a machine, and the universe was not comprehended "as a as matter undergoing uninterrupted historical development". Relatively autonomous objective systems with higher forms of motion were reduced to mechanical ones.

Marx and Engels, as well as Hegel (the latter remained trapped in irrational thinking, although he revolutionised philosophical methodology), were highly critical of the Newtonian worldview. They emphasized interconnection and processes instead of singularities and reduction. Hegel criticized atomistic philosophies by saying that they fix the One as One, "the Absolute is formulated as Being-for-self, as One, and many ones." The One and the Many are dialectically connected: the One is being-for-itself and related to itself, but this relationship only exists in relationship to others (being-for-another), and hence it is one of the Many and repulses itself. "But the Many are one the same as another: each is One, or even one of the Many; they are consequently one and the same.... As those to which the One is related in its act of repulsion are ones, it is in them thrown into relation with itself. The repulsion therefore has an equal right to be called Attraction; and the exclusive One, or Being-for-self, suppresses itself".

Marx and Engels, in criticizing Max Stirner's reductionism and individualism, put forward the notion of the individual as a social being that is estranged in capitalism and can only become a well-rounded individual in communism (1976, 117-427). Engels criticized the reductionism and individualism of "metaphysical thinkers":

To the metaphysician, things and their mental reflexes, ideas, are isolated, are to be considered one after the other and apart from each other, are objects of investigation fixed, rigid, given once for all. He thinks in absolutely irreconcilable antitheses. "His communication is 'yea, yea; nay, nay,' for whatsoever is more than these cometh of evil" [Matthew 5:37—Ed.]. For him a thing either exists or does not exist; a thing cannot at the same time be itself and something else. Positive and negative absolutely exclude one another, cause and effect stand in a rigid antithesis one to the other. (1976a, 22)

Hard And Fast Lines are incompatible with the theory of evolution.... For a stage in the outlook on nature where all differences become merged in intermediate steps, and all opposites pass into one another through intermediate links, the old metaphysical method of thought no longer suffices. Dialectics, which likewise knows no HARD AND FAST LINES, no unconditional, universally valid "either—or" and which bridges the fixed metaphysical differences, and besides "either—or" recognises also in the right place "both this—and that" and reconciles the opposites, is the sole method of thought appropriate in the highest degree to this stage.

Self-organisation theory today also stresses the inter-connectedness and process-structure of the world and criticizes reductionism. Ilya Prigogine and Isabelle Stengers, the founders of dissipative systems theory, stress that Hegel, Marx, and Engels are important process-thinkers in this regard: "The Hegelian philosophy of nature systematically incorporates all that is denied by Newtonian science. In particular, it rests on the qualitative difference between the simple bahaviour described by mechanics and the bahaviour of more complex entities such as living beings. It denies the possibility of reducing those levels, rejecting the idea that differences are merely apparent and that nature is basically homogeneous and simple" (1984, 89). "The idea of a history of nature as an integral part of materialism was asserted by Marx and, in greater detail, by Engels. Contemporary developments in physics, the discovery of the constructive role played by irreversibility, have thus raised within the natural sciences a question that has long been asked by materialists. For them, understanding nature meant understanding it as being capable of producing man and his societies".

Marx and Engels opposed the idea of substance (an everlasting, changeless carrier of changing qualities as primary matter because they considered such a position as mechanical and undialectical, and argued that it neglected the fact that matter is always in motion, and develops higher levels of organisation in the dialectical process of becoming. In the history of dialectical materialism, one finds an animosity toward the notion of substance. Lenin, for example, wrote: "The recognition of immutable elements, 'of the immutable substance of things,' and so forth, is not materialism, but metaphysical, i.e., anti-dialectical, materialism". Herbert Horz, one of the main philosophers of the German Democratic Republic, argued that owing to the physics of fields, the discovery of radioactivity, relativity theory, and quantum theory, the notion of substance has become untenable. Modern physics has shown that elementary particles are transformed into one another; particles arise and continue to exist only in qualitative relationships to others. Hence the idea of an unchangeable carrier of qualities seems no longer to be valid. "Whereas the substance concept presupposes an unchanging carrier,... modern physics conceives material events primarily as change and interaction, and searches for the structural laws of this change". The notion of substance would not be able to show the dialectical relationship of particle and field that was introduced by quantum theory. Fields and elementary particles cannot be substance because they are subject to change.

Hegel opposed the notion of substance for other reasons: Spinoza sees substance as causa sui—it is its own reason. Hegel says that such an assumption would exclude the creation of the world by God, which he believed in. "A deeper insight into nature reveals God as creating the world out of nothing. And that teaches two things. On the one hand it enunciates that matter, as such, has no independent subsistence, and on the other that the form does not supervene upon matter from without, but as a totality involves the principle of matter in itself".

Modern physics repudiates the mechanistic and reductionist conception of substance. Nonetheless, there seems to be an alternative conception of substance immanent in Engels's works on nature: The substance of the world—that which exists permanently and endlessly—is the process-structure of matter. Matter is without pause in permanent motion, in ceaseless flux, and is a self-producing entity. In its dialectical movement it produces different levels of organisation that have higher, emergent qualities that cannot be reduced to earlier qualities. Engels stressed that matter is a producing entity, and through its permanent flux and motion "remains eternally the same in all its transformations".

The Marxist philosopher Ernst Bloch worked out an alternative conception of substance and matter within the framework of dialectical materialism. In opposition to mechanical materialism, Bloch argues that matter is process-like; it is not a "dead block, moved only by pressure and push and remaining itself all the time" (1963, 230), but nonetheless he does not give up the notion of substance. Matter for Bloch is fermenting and process-like (203); it is a process-being, being-in-possibility (1963, 207) and has a historical-dialectical character (209). Bloch's concept of matter anticipated the modern theories of self-organisation that also stress the productivity of matter resulting in different organisational forms and hierarchical levels of matter and the self-reproduction and recreativity of self-organising units.

Nature is for Bloch a producing subject; he says it is forming itself, forming out of itself (234). In this context Bloch takes up Spinoza's concept of natura naturans in order to stress that nature is not only passively produced, but is also itself an actively producing system. The relationship of tendency and latency in matter also reappears as a dialectic of chance and necessity in self-organisation theory (the concepts of relative chance by Kolmogorow and Chaitin and of incomplete determinism). What Bloch calls a novum is called emergent qualities in the sciences of complexity. Bloch used the term emergence himself by stressing that all gestalt figures emerge from the dialectical process and from matter as developing, producing substance immanently as well as speculatively. For Bloch matter is a dialectically developing, producing substance. Substance for Bloch is process-substance (1975, 246); it opens up possibilities, is fermenting, and actively producing.

Self-organisation and Dialectics

Saying the substance of the world is the permanent dialectical movement of matter and its self-productivity corresponds to saying that matter organises itself and that nature is a self-organising system. Wolfgang Hofkirchner has stressed that the new results of scientific research have been anticipated by Marx and Engels, and that the concept of dialectical development reenters science with self-organisation theory.

The theory of self-organisation has led to a change of scientific paradigms—from the Newtonian paradigm to the approaches of complexity. There is a shift from predictability to nonpredictability; from order and stability to instability, chaos, and dynamics; from certainty and determination to risk, ambiguity, and uncertainty; from control and steering to the self-organisation of systems; from linearity to complexity and multidimensional causality; from reductionism to emergentism; from being to becoming; and from fragmentation to interdisciplinarity. This has been interpreted as a shift from modern to postmodern knowledge.

Concepts of physical self-organisation have been put forward in Ilya Prigogine's theory of dissipative systems, Hermann Haken's synergetics (1978, 1983), and Manfred Eigen's hypercycle theory.

The principles of physical self-organisation are:

1. Control parameters: A set of parameters influences the state and bahaviour of the system.
2. Critical values: If certain critical values of the control parameters are reached, structural change takes place and the system enters a phase of instability/criticality.
3. Fluctuation and intensification: Small disturbances from inside the system intensify themselves and initiate the formation of order.
4. Feedback loops, circular causality: Feedback loops occur within a self-organising system; circular causality involves a number of processes p1, p2,..., pn (n ¡Ý 1), and p1 results in p2, P2 in p3,..., pn-1 in pn and pn in p1.
5. Nonlinearity: In a critical phase of a self-organising system, causes and effects cannot be mapped linearly: similar causes can have different effects and different causes similar effects; small changes of causes can have large effects, whereas large changes can also result in only small effects (but nonetheless it can also be the case that small causes have small effects and large causes, large effects).
6. Bifurcation points: Once a fluctuation intensifies itself, the system enters a critical phase where its development is relatively open, certain possible paths of development emerge, and the system has to make a choice. This means a dialectic of necessity and chance. Bifurcation means a phase transition from stability to instability.
7. Selection: In a critical phase that can also be called point of bifurcation, a selection is made between one of several alternative paths of development.

8. Emergence of order: In a critical phase, new qualities of a self-organising system emerge; this principle is also called order from chaos or order through fluctuation. A self-organising system is more than the sum of its parts. The qualities that result from temporal and spatial differentiation of a system are not reducible to the properties of the components of the systems; interactions between the components result in new properties of the system that cannot be fully predicted and cannot be found in the qualities of the components. Microscopic interactions result in new qualities on the macroscopic level of the system. Checkland defines an emergent quality in similar terms "as a whole entity which derives from its component activities and their structure, but cannot be reduced to them" (1981, 314). The emergence of order includes both (a) bottom-up-emergence (a perturbation causes the system's parts to interact synergetically in such a way that at least one new quality on a higher level emerges) and (b) downward causation (once new qualities of a system have emerged they, along with the other structural macro aspects of the system, influence—that is, enable and constrain—the bahaviour of the system's parts). This process can be described as top-down-emergence if new qualities of certain parts (seen as wholes or systems themselves) show up.
9. Information production: Information is a relationship between specific organisational units of matter. Reflection means reaction to influences from the outside of a system in the form of inner-systemic structural changes. A causal relationship exists between the result of reflection and the reflected. The reflected causes structural changes, but does not mechanically determine them. The system has a certain relative autonomy that can be described as a degree of freedom from perturbations. On the different organisational levels of matter, we find different degrees of freedom. The degree increases along with complexity if we go up the hierarchy from physical-chemical to living systems, and finally to social systems. The causal relationship between the reflected and the result of reflection is based on a dialectical relationship of freedom and necessity. Information is an objective relationship between the reflected, the result of reflection inside the system's structure, and the realisation of functions of the system within the reflected environment of

the system. This means that information is a relationship of creative and active reflection between a system and its environment—to be more precise, between units of organised matter. Stimuli and fluctuations cause inner-systemic structural change; fluctuations are actively reflected within the system. Information is not a structure given in advance; it is produced within material relationships. "Information is a physical structure and at the same time a structure which dominates the physical forces.... Information is not a physical substance; it is instead temporarily 'attached' to it. Information must be understood as a specific effect and as a relationship".

10. Fault tolerance: Outside a critical phase, the structure of the system is relatively stable concerning local disturbances and a change of boundary conditions.
11. Openness: Self-organisation can only take place if the system imports entropy that is transformed; as a result, energy is exported, or, as Prigogine says, dissipated.
12. Symmetry breaking: The emerging structures have less symmetry than the foundational laws of the system.
13. Inner conditionality: Self-organising systems are influenced by their inner conditions and the boundary conditions from their environment.
14. Relative chance: There is a dialectic of chance and necessity in self-organising systems; certain aspects are determined, whereas others are relatively open and subject to chance.
15. Complexity: Self-organising systems are complex systems. The term complexity has three levels of meaning: (1) There is self-organisation and emergence in complex systems. (2) Complex systems are not organised centrally, but in a distributed manner; there are many connections between the system's parts. (3) It is difficult to model complex systems and to predict their bahaviour even if one knows to a large extent the parts of such systems and the connections between the parts. The complexity of a system depends on the number of its elements and the connections between the elements (the system's structure). According to this assumption, Kauffman defines complexity as the "number of conflicting constraints" in a system (1993). Heylighen says that complexity can be characterised by a lack of symmetry (symmetry breaking),

which means that "no part or aspect of a complex entity can provide sufficient information to actually or statistically predict the properties of the others parts" (1996), and Edmonds defines complexity as "that property of a language expression that makes it difficult to formulate its overall bahaviour, even when given almost complete information about its atomic components and their interrelations" (1999). Aspects of complexity are things, people, number of elements, number of relations, nonlinearity, broken symmetry, nonholonic constraints, hierarchy, and emergence.

16. Cohesion: Cohesion means the closure of the causal relations among the dynamical parts of a dynamical particular that determine its resistance to external and internal fluctuations that might disrupt its integrity. It is a "dividing glue" of dynamic entities.
17. Systemness: Self-organisation takes place in a system—in a coherent whole that has parts, interactions, structural relationships, bahaviour, state, and a border that delimits it from its environment.
18. Hierarchy: The self-organisation of complex systems produces a hierarchy in two distinctive senses: (1) The level of emergence is a hierarchically higher level—that is, it has additional, new emergent qualities that cannot be found on the lower level that contains the components. The upper level is a sublation of the lower level. (2) Self-organisation results in an evolutionary hierarchy of different system types; these types are hierarchically ordered in the sense that upper levels are more complex and have additional emergent qualities.
19. Globalisation and localisation: Bottom-up-emergence means the globalising sublation of local entities; downward causation means the localisation of more global qualities.
20. Unity in plurality (generality and specificity): The organising system is characterised by a number of distinctive qualities that distinguish it from other self-organising systems. On the other hand, each type of self-organising system also shares general principles and qualities with all other types of self-organising systems. Both generality/unity and specificity/plurality are characteristic of self-organising systems.

The concept of emergence is the central notion of self-organisation concepts. Aspects of emergence are:

Synergism: Emergence is due to the productive interaction between entities. Synergy is a very general concept that refers "to combined or 'co-operative' effects—literally, the effects produced by things that 'operate together' (parts, elements or individuals)". Synergy takes place and shapes systems on all organisational levels of matter; it is a fundamental quality of matter. Synergies between interacting entities are the cause of the evolution and persistence of emergent systems.

Novelty: On a systemic level different from the level of the synergetically interacting entities, new qualities show up. Emergent qualities are qualities that have not been previously observed and have not previously existed in a complex system ("a whole is more than the sum of its parts").

Irreducibility: The newly produced qualities are not reducible to, or derivable from, the level of the producing, interacting entities.

Unpredictability: The form of the emergent result and the point of emergence cannot be fully predicted.

Coherence/correlation: Complex systems with emergent qualities have some coherent bahaviour for a certain period of time. This coherence spans and correlates the level of the producing entities into a unity on the level of emergence. Historicity: Emergent qualities are not given a priori, but are the result of the dynamical development of complex systems.

One example of physical self-organisation is the Benard cells: A special liquid is heated at a certain temperature t2 from beneath and cooled down to a certain temperature t1 from above. The temperature difference ¦£t = t2 - t1 becomes control parameter of the system (principle 1). At ¦£t = 0 the system is in equilibrium, the temperature gradient rises, and, at a certain critical value (principle 2), a new pattern emerges in the liquid that looks like honeycombs (principles 8 and 9). The liquid particles are located in layers; lower layers result from their temperature being warmer than upper ones; they expand and their density decreases. At the beginning of the critical phase, a first small fluctuation occurs, which means that a particle is displaced from its position in a certain layer and enters an upper or lower layer (principle 3). The layer in which this fluctuation will occur is not predetermined. Fluctuations only take place if a certain threshold of the control parameter At is crossed. The fluctuation intensifies itself (principle 3); more and more liquid particles are detached from their stationary position; disorder, chaos, and motion appear (principle 6).

The liquid particles arrange themselves in cells that have different forms (round, square, broad, thin, large, small, etc.). These forms are dependent on the elementary modes of motion. Several types of cells exist simultaneously. Finally, one type can assert itself and become a dominant form due to a selection process within the system (principle 7). As a result of the superimposition of many cells of the same form, a pattern emerges that looks like a honeycomb (principles 8 and 9). So from an initial chaos of particles, order has emerged. At a certain value of the temperature gradient, this order disappears. In this process, order will definitely emerge, initial fluctuations will spread out, and one of several types of roles will be selected. But it is not determined in which layer the fluctuation will occur, exactly how the cell-types will look, and which one will be selected (principle 14). This experiment will only be successful if energy in the form of a temperature difference is applied to the system (principle 11).

A laser is another example that is frequently used to explain self-organisation. A laser consists of an active medium that is situated between two mirrors. This medium is either a gas that is radiating due to the discharge caused by the entry of current or a crystal that is pumped by a flash lamp (e.g., a ruby with chrome ions). The flashes stimulate the crystal, and an electron changes its trajectory, jumping from an inner trajectory to an outer one, absorbing energy from the flash lamp. It spontaneously returns to its former trajectory and emits energy in the form of a light wave. Therefore, the atoms emit light waves because of their stimulation by the flash lamp. The two mirrors continually reflect the light. First there is a chaos of light waves. A light wave can hit other atoms and force an increase in the light intensity. By such processes, the light waves reach certain amplitudes. Haken says that one light wave "enslaves" the others; this means that it becomes dominant and orders the system. As a result, an ordered light wave—the laser beam—emerges. From a chaos of light waves, an ordered pattern emerges (principles 8 and 9). The decisive control parameter is the current supply (principles 1 and 11). The system can only enter criticality if the current reaches a certain threshold (principle 2). A light wave is caused by a fluctuation—that is, an electron returns to its inner trajectory and emits energy; a light wave can intensify itself by "enslaving" electrons (principle 3). Such an intensification always means circular causality, because one entity causes the bahaviour of another entity, and this bahaviour results in a transformation of the first entity (principle 4). Due to such intensifications, the system enters a state of chaos/instability/ bifurcation (principles 5 and 6). A certain light wave is selected (principle 7) and determines the emergence of the laser beam (principles 8 and 9). It is determined that a laser beam will emerge, that fluctuations and intensification will result; but not determined is exactly how this will take place and which light wave will order the system (principle 14).

According to Hegel's outline, the purpose of dialectics is "to study things in their own being and movement and thus to demonstrate the finitude of the partial categories of understanding" (1973, note to §81). Self-organisation refers to the forms of movement of matter and hence is connected to dialectics. What are called control parameters, critical values, bifurcation points, phase transitions, nonlinearity, selection, fluctuation, and intensification in self-organisation theory (principles 1, 2, 3, 5, 6, 7) correspond to the dialectical principle of transition from quantity to quality. This corresponds to what Hegel discussed as Measure: Measure is the qualitative quantum—quantum is the existence of quantity.

The identity between quantity and quality, which is found in Measure, is at first only implicit, and not yet explicitly realised. In other words, these two categories, which unite in Measure, each claim an independent authority. On the one hand, the quantitative features of existence may be altered, without affecting its quality. On the other hand, this increase and diminution, immaterial though it be, has its limit, by exceeding which the quality suffers change....

... But if the quantity present in measure exceeds a certain limit, the quality corresponding to it is also put in abeyance. This however is not a negation of quality altogether, but only of this definite quality, the place of which is at once occupied by another. This process of measure, which appears alternately as a mere change in quantity, and then as a sudden revulsion of quantity into quality, may be envisaged under the figure of a nodal (knotted) line.

What is called emergence of order, production of information or symmetry breaking in self-organisation theory corresponds to Hegel's notions of sublation and negation of the negation. Something is only what it is in its relationship to another, but by the negation of the negation this something incorporates the other into itself. The dialectical movement involves two moments that negate each other, a somewhat and an another. As a result of the negation of the negation, "something becomes an other; this other is itself somewhat; therefore it likewise becomes an other, and so on ad infinitum". Being-for-self or the negation of the negation means that somewhat becomes an other, but this again is a new somewhat that is opposed to an other and as a synthesis results again in an other and therefore it follows that something in its passage into other only joins with itself; it is self-related. In becoming, there are two moments: coming-to-be and ceasing-to-be: by sublation—being passes over into nothing; it ceases to be, but something new shows up, is coming to be. What is sublated on the one hand ceases to be and is put to an end, but on the other hand, it is preserved and maintained. In dialectics, a totality transforms itself; it is self-related. This corresponds to the notions of self-production and circular causality. The negation of the negation has positive results—that is, in a self-organising system, the negation of elements results in positive new qualities.

The two examples mentioned here in fact are examples of the dialectical development of matter. Hegel says that when the control parameters reach a certain threshold, a point of bifurcation, or criticality, a nodal-line arises. The quantities that are increased and transform into quality are the temperature gradient and electric current. The emergence of the honeycomb pattern or the laser beam means sublation and negation of the negation. The old states of the systems are eliminated, but nonetheless preserved in new qualities. New qualities arise and the systems thereby reach a higher level.

The principle of relative chance that is typical for self-organising systems had already been considered as a dialectic of chance and necessity by Hegel, Marx, and Engels. Engels stressed that the dialectic of attraction and repulsion is an aspect of matter and its movement. Both elements are also described by self-organisation theory: chaos, noise, or instability is described as disordered movement of the elements of a complex system. One can also say that the elements are repelling each other. But this repulsion is one that turns into attraction, because the elements interact, there are processes of ordering and selection—attraction takes place as the emergence of coherent whole and new qualities. As an example of the transition from quantity to quality, Engels mentions the homologous series of carbon compounds:

Here therefore we have a whole series of qualitatively different bodies, formed by the simple quantitative addition of elements, and in fact always in the same proportion. This is most clearly evident in cases where the quantity of all the elements of the compound changes in the same proportion. Thus, in the normal paraffins CnH2n+2, the lowest is methane, CH4, a gas; the highest known, hexadecane, C16H34, is a solid body forming colourless crystals which melts at 21° and boils only at 278°. Each new member of both series comes into existence through the addition of CH2, one atom of carbon and two atoms of hydrogen, to the molecular formula of the preceding member, and this quantitative change in the molecular formula produces each time a qualitatively different body.

Nodal lines or the transition from quantity to quality is today also studied in self-organisation theory. The theory of self-organised criticality (SOC) especially focuses on this. It studies phenomena where perturbations that normally have small effects have large effects in a critical situation and push the system into chaos. A frequently mentioned example is a pile of sand. Dropping grains of sand onto each other will result in a pyramid. When the pile reaches a certain critical point, there is the possibility that just one additional grain results in the avalanching collapse of the whole pile. In a phase of SOC, the effects of one additional element vary from small to large, either pushing the system into chaotic bahaviour or locking it into a fixed bahaviour. The system is on the "edge of chaos." One feature that characterises SOC systems is a power-law distribution of the characteristic events such as avalanches, quakes, crashes, etc. The average frequency of the event is inversely proportional to some power of its size: log (F) = -log (M). The log of the frequency of events is a linear function of the log of their magnitudes. The theory of SOC assumes that SOC patterns can be found, for example, in wars, wildfires, stock prices, traffic jams, international conflicts, and the collapse of society.

Almost everywhere in chemistry one can find examples of the transition from quantity to quality. Therefore Engels speaks of chemistry as the "science of the qualitative changes of bodies as a result of changed quantitative composition". This transition is what today in self-organisation theory is called emergence. 7 In a self-organising system, a certain threshold of a control parameter is crossed and order emerges. What is today called a point of bifurcation, instability, or criticality, Engels refers to as "Hegelian nodal line of measure relations—in which quantitative change suddenly passes at certain points into qualitative transformation", or even directly anticipating the modern terminology, he speaks of "critical point". As other examples of nodal lines, Engels mentions a certain current strength that is required to cause the platinum wire of an electric incandescent lamp to glow, the temperatures of incandescence and fusion of metals, the freezing and boiling points of liquids, the critical point at which a gas can be liquefied by pressure and cooling.

The transition from quantity to quality that occurs, for example in the homologous series of carbon compounds when certain atoms are added can also be termed the emergence of a qualitatively different body.

Other examples that Engels mentioned for the transition from quantity to quality, and that could equally be described as the emergence of new qualities in a critical situation after a threshold of a certain control parameter has been crossed, include:

Change of Form of Motion and Energy

All qualitative differences in nature rest on differences of chemical composition or on different quantities or forms of motion (energy) or, as is almost always the case, on both. Hence it is impossible to alter the quality of a body without addition or subtraction of matter or motion, i.e. without quantitative alteration of the body concerned....

Change of form of motion is always a process that takes place between at least two bodies, of which one loses a definite quantity of motion of one quality (e.g., heat), while the other gains a corresponding quantity of motion of another quality (mechanical motion, electricity, chemical decomposition). Here, therefore, quantity and quality mutually correspond to each other. (1987b, 357)

Engels's Citation of Hegel's Example of the States of Aggregation of Water

Thus the temperature of water is, in the first place, a point of no consequence in respect to its liquidity: still with the increase of diminution of the temperature of the liquid water, there comes a point where this state of cohesion suffers a qualitative change, and the water is converted into steam or ice.

As other examples, Hegel mentions that a point is reached where a single additional grain makes a heap of wheat; or where a bald tail is produced by plucking a single hair from a horses tail.

For Engels, "the negation of the negation" is "an extremely general ... law of development of nature, history, and thought; a law which, as we have seen, holds good in the animal and plant kingdoms, in geology, in mathematics, in history and in philosophy" (1987a, 131), As an example from nature, he mentions the development process of a grain of barley: Billions of such grains of barley are milled, boiled and brewed and then consumed. But if such a grain of barley meets with conditions which are normal for it, if it falls on suitable soil, then under the influence of heat and moisture it undergoes a specific change, it germinates; the grain as such ceases to exist, it is negated, and in its place appears the plant which has arisen from it, the negation of the grain. But what is the normal life-process of this plant? It grows, flowers, is fertilised and finally once more produces grains of barley, and as soon as these have ripened the stalk dies, is in its turn negated. As a result of this negation of the negation we have once again the original grain of barley, but not as a single unit, but ten-, twenty- or thirtyfold.

As similar examples, he mentions the development process of insects, geology as a series of negated negations, a series of successive shatterings of old and deposits of new rock formations, differential and integral calculus, the development of philosophy and society. These development processes can also be described in terms of physical self-organisation: the control parameters that influence the development of the grain are time and natural conditions such as heat and moisture. During this development, new seeds will show up. At a certain time, a critical point is reached and the grain ceases to exist. But at the same time, new grains emerge.

Dialectical processes and negation of the negation mean not only just the emergence of other, new qualities. Dialectical development also includes development process that results in higher qualities and other structural levels. Dialectical development is not just change or self-transformation and self-reproduction; it is also the emergence of higher levels of organisation. Hence dialectical thinking assumes an immanent hierarchy in nature and evolutionary leaps. This was also pointed out by Engels:

The transition from one form of motion to another always remains a leap, a decisive change. This is true of the transition from the mechanics of celestial bodies to that of smaller masses on a particular celestial body; it is equally true of the transition from the mechanics of masses to the mechanics of molecules—including the forms of motion investigated in physics proper: heat, light, electricity, magnetism. In the same way, the transition from the physics of molecules to the physics of atoms—chemistry—in turn involves a decided leap; and this is even more clearly the case in the transition from ordinary chemical action to the chemism of albumen [proteins] which we call life. Then within the sphere of life the leaps become ever more infrequent and imperceptible.

Self-organisation theory is also dialectical in the respect that it frequently considers self-organisation as emergent evolution. This means that there are different hierarchical organisational levels of self-organisation that differ in complexity and where new qualities of organisation emerge on upper levels. In self-organisation theory, Ervin Laszlo, for example, argues that evolution does not take place continuously, but in sudden, discontinuous leaps (1987). After a phase of stability, a system enters a phase instability, fluctuations intensify and spread out. In this chaotic state, the development of the system is not determined; what is determined is only that one of several possible alternatives will be realised. Laszlo says that evolution takes place in such a way that new organisational levels emerge, constituting the successive steps of evolution. Not all scientists who speak about self-organisation include the development of higher qualities into their concepts. Hence, in this respect, dialectical materialism can be considered as a broader evolutionary concept than self-organisation.

In his Anti-Duhring and his Dialectics of Nature, Engels pointed out the problem of defining life and intuitively anticipated the theory of autopoiesis. Of course today we know much more about life than Engels did, especially since the discovery of the double helix. But what is important is that Engels anticipated the idea of autopoiesis. He said that life exists in the "constant self-renewal of [its own] chemical constituents"; life is a "self-implementing process". Proteins not only continually undergo decomposition, but also continually produce themselves from their components.

Science, Materialism, and Religion

As Engels implicitly pointed out, the substance of the world is its process character, the continual dialectical movement of matter, and the productivity of matter that results in self-reproduction and the emergence of new, higher qualities and organisational forms of matter. This corresponds to saying that the substance of the world is the continual self-organisation of matter. As has been shown, processes of physical self-organisation can be described in dialectical terms. Control parameters, critical values, bifurcation points, phase transitions, nonlinearity, selection, fluctuation, and intensification in self-organisation theory correspond to the dialectical principle of transition from quantity to quality. What is called emergence of order, production of information, or symmetry breaking in self-organisation theory corresponds to Hegel's notions of sublation and negation of the negation. The concept of emergent evolution corresponds to the principle of dialectical development, the dialectics of chance and necessity, as well as of attraction and repulsion that have been described by Hegel, Engels, and Marx are constitutive for processes of self-organisation. Conversely, the examples Engels gave for the dialectics of nature can also be seen as examples of the self-organisation of matter.

Self-organisation theory shows that Engels's Dialectics of Nature is still very topical and that dialectical materialism, contrary to mechanical materialism, has not been invalidated; rather, it confirms that dialectics is the general principle of nature and society. Self-organisation theory supports Engels's assumptions that the real unity of the world consists in its materiality, that matter is process-like and in constant flux, that it is a producing entity that is uncreateable and indestructible. That the substance of the world is self-organisation of matter, which results in higher forms of organisation of matter the highest form of organisation of matter thus far being human society—means that God does not exist, that there is no creatioexnihilo and no first mover that is not itself moved. Hence religion and esoteric thinking are mere ideology and false consciousness. Dialectical materialism seems to be confirmed by modern science, whereas serious problems arise for idealist worldviews. "The conceptions of self-organisation, the conceptions that assign a determining role to the activity of inner factors instead of outer, are new scientific affirmations of the old dialectical theses, as well as the conceptions of the general connection of all things and appearances". Self-organisation theory is indeed a dialectical-materialist theory, but unfortunately its representatives all too often do not realise this and do not acknowledge the dialectical tradition and heritage of the philosophy of nature in the line of Frederick Engels and Karl Marx.

The natural sciences that emerged during the last century, such as quantum theory, quantum mechanics, first- and second-order cybernetics, general system theory, nonequiblibrium thermodynamics, synergetics, dissipative systems theory, autopoietic systems theory, catastrophe theory, punctuated equilibrium theory, hypercycle theory, string theory, loop theory, etc. deal with the ontology of the material world. Hence there seems to be scientific evidence that nature is a self-organising totality and is its own cause. This seems to confirm the materialist notion that matter is uncreateable and indestructible.

Twentieth-century science indicates that dialectical development is a universal law of nature and that dialectical materialism is correct, but that human consciousness frequently lags behind the progress of science, technology, and society. Linked to the current crisis of the capitalist world system is a tendency to spread mysticism and irrationalism in society. This tendency also affects the scientific community. It is quite common today in idealistic thinking to interpret the big bang as the creation of the world by God, where nothing turns into something. But if before the big bang there was nothing except God, what is the foundation of God? There has never been scientific evidence that God could really exist as an eternal substance outside of material existence and that God is his or her own reason, whereas modern science has produced evidence that matter is causa sui, organises itself and has not been created by an external first mover out of nothing. It is not reasonable to assume that the world has been created out of nothing by God and that God really exists. In such arguments, a causal principle is applied to matter, but the same causal principle is declared as not holding for God. There are no rational reasons why this should be the case. Talking about God and the origin of the world means talking about universality. It is unreasonable to apply a form of universal causality to one universal phenomenon, but simply ignore it for another one.

Philosophy deals with explanations of how single aspects of the world and single sciences are connected. It is the science of universality.

Philosophy is the thinking study of material reality and the things that comprise reality. Philosophy works out notions and categories in order to describe and explain the total world process on a general level. Various idealistic, religious, and esoteric theories explain the world as being created by God as an external first mover who is not moved himself. This violates fundamental philosophical theorems such as Occam's Razor: if the material world can be explained as its own reason, as can be done by philosophically generalising theories of self-organisation, reference to an external creator is an unnecessary over-specification and multiplication. The theorem of foundation holds that everything that is or can be has some foundation or ground. With physics serving as the starting point for the history of the cosmos, matter can be conceived as its own reason and as the self-referential foundation of the world. Philosophy actually must explain the development of the universe, and must start from physics as the fundamental natural science; idealistic conceptions that stress spirit will fail to find a sufficient ground of the universe. If Spirit and God are conceived as eternal entities that are their own reason, irrational categories are simply defined tautologically and without reference to the really existing, material world that can be rationally explained by the natural and social sciences. Idealism cannot provide a reasonable foundation of the world.

While we have no scientific proof for the existence of God, we have every reason to assume that matter is organising itself and that this is a universal phenomenon. Manfred Eigen's hyper-cycle theory provides an explanation of the origin of life and the human being that requires no argument assuming divine creation, because it explains the emergence of life as a qualitative leap in the self-organisation of matter that results in a new level of organisation within an evolutionary hierarchy. Life is the result of a cross-catalysis between autocreative nucleic acids and proteins. "There is no need for a miracle, for a divine, supernatural act to explain biological development. The only possibility of avoiding this conclusion would be the statement that the laws ruling it have been created together with the world by an extra-human force. But then reasonable arguments for the possibility and necessity of this extranatural power must be found, and that cannot be established by scientific means". The existence of life is due to self-reproducing molecules; there is no scientific evidence for a creation of life and human beings by God. In one of its versions, idealism is based on a dualism of mind and matter; in another, matter is reduced to mind. Examination of the history of the division of labour shows that this division resulted in a widening separation between manual and mental labour. The emergence of this separation coincides with the emergence of class-based society. Idealism received a boost from the emergence of classes and heteronomous societies; conversely, it is an ideology that justifies and is helpful in upholding such societies.

With the breakdown of Fordist capitalism in the sixties, the capitalist world system entered a permanent crisis, and global problems have quickly worsened since then. A new post-Fordist mode of capitalist development emerged and individualisation has shown up as a new phenomenon that serves dominating interests and results in the erosion of collective institutions that formerly seemed to give sense to the human being. Such institutions are traditional religions, unions, associations, families, etc. Capitalism is now based on a deregularised and flexible institutional setting (flexible regime of accumulation, neoliberal mode of regulation), and people throughout the world are faced with the dangers of precariousness and extinction due to the development of the internal antagonisms of the capitalist world system. With the breakdown of the Soviet Union, an ideological vacancy appeared, and the former Eastern European states have been fully included into the global capitalist dynamics.

In ideology and science, the emptiness and helplessness felt by many due to the antagonisms of the capitalist world system have resulted in a search for new transcendental and mystical explanations and salvations. As a result, there is a boom of various forms of mysticism, esotericism, and spiritualism. People are looking for irrational guidelines, instead of looking for the foundations of problems and developments within the real world. The new irrationalism is a result of the increased complexity of the world with which people cannot cope.

These irrational tendencies can also be found within the self-organisation paradigm that has been interpreted by some as holistic spiritualism.

In such mystical views, the universe is seen as one large living totality that consists of a network of equal parts. There is no hierarchy in nature in such conceptions and hence also no qualitative differences between systems. They are all considered as an expression of spirit. Based on the Gaia hypothesis, biologistic and ecofascistic arguments are frequently employed. In such new mystifications and irrational understandings of science, God is not necessarily considered as an eternal creator, but an eternal principle exists external to matter. Philosophy is not an area of religious belief; religion is not a part of science and philosophy. Values and norms are part of ethics, which comprises one part of philosophy. The other parts are ontology (what is the world and all being like?) and epistemology (how do we perceive the world?). Philosophy is not an area where "anything goes" in the sense of a radical constructivist or anarchistic epistemology of science as put forward, for example, by Paul Feyerabend. Philosophy, instead, tries to connect, to generalise, and to unify single sciences. It produces interrelationships between single sciences on a more general metalevel. Hence it is based on the natural and social sciences; philosophical categories are related to the single sciences; categories like reason, love, human being are related to the humanities; categories like nature, space, time, matter are related to physics, etc.

Categories like God and Spirit that are conceived as the Absolute, as something infinite and unquestionable and as absolute truth, are not at all connected to the single sciences. This results in isolated doctrines that cannot be analysed, questioned, and examined scientifically. For example, there is no proof for the claim that humans occupy some lower steps in a universal field where God means the Absolute. The realm of religion, mysticism, spiritualism, and esotericism is where science ends and pure ideology starts.

Hegel said that "what is reasonable is actual and what is actual is reasonable." Actuality means materiality, hence turning Hegel right side up means that only material reality can be reasonable, and that something that is conceived as existing prior or external to matter is unreasonable. Areas such as religion and esotericism are unscientific and irrational; they proclaim absolute truths that cannot be researched or contested. Irrational arguments avoid objectivity, exactness, logic, verifiability, and falsifiability. Pseudosciences use strategies of immunisation in order to avoid criticism. If pseudosciences like creationism, spiritualism, mysticism, parapsychology, and astrology were right, this would mean that the modern sciences are all wrong. Hence isolationism is typical for such areas of thinking.

Religion and other irrationalisms have no scientific grounds. Religions might include some elements that are interesting for science and philosophy, but one should deal with these topics scientifically, not religiously and in terms of absolute truths. Religion and esoterics are a "universal basis of consolation and justification.... Religious distress is at the same time the expression of real distress and also the protest against real distress. Religion is the sigh of the oppressed creature, the heart of a heartless world, just as it is the spirit of spiritless conditions. It is the opium of the people".

There is no need to refer to mystic forces for explaining the self-organisation of the universe and society. New properties simply emerge due to the complex interactions of the parts of a system, not because some external holistic force is at play. The founders of the philosophy of emergentism, Conwy Llord Morgan and Samuel Alexander, saw emergence as something mystic, and so they introduced spiritual forces (known as "Nisus") as the driving principle. To posit such forces shows a lack of understanding of the dialectical relationship of quality and quantity and the whole and its parts. The emergence of order need not be explained metaphysically, because new qualities of the whole are solely constituted by interactions of its parts. The philosophical mistake of overspecification that is grasped by Occam's razor is made by holistic thinkers such as Jantsch and Capra. This opens the way for irrationalism and esotericism, which belong to the scope of ideology rather than to (critical) science.

The Strategic Implications

Three science-based technological revolutions are currently under way: one is in materials science and engineering, another is in the biosciences and biotechnology, and the third is in information technologies and the convergence of computers, communications, and multimedia content. These advances are multidisciplinary, largely interdependent (e.g., bioinformatics, biomaterials, and computer-aided modelling and simulation of advanced structural materials, or of molecular structures in the drug discovery process), and are proceeding at a rapid pace from local networks and research centres of excellence located within different national systems of innovation (NSI) around the world. Moreover, current developments under way in the field of nanotechnology involve the convergence of the physical, chemical, and biological sciences and will therefore usher in a new era of full integration of the scientific and engineering fields, underpinning advances in all three fields mentioned above, over the next ten years.

This chapter is based on ongoing original research and illustrative case studies, some of which were first reported in the Financial Times. Its aim is to offer new theoretical and empirical insights and propositions regarding the conceptualisation of emerging scientific and engineering revolutions and their implications for corporate technology strategy and the global location of production and R&D activities. The analysis is conducted in the context of the debate between technoglobalism versus technonationalism and of the trends toward the development of the knowledge-based economy and of the digitized or inter-networked world economy. The chapter offers a first examination of, and some initial empirical evidence on, the following set of hypotheses, which form the basis for further research and conceptual elaboration:

Hypothesis 1: The developments under way in the three science-based generic technologies, and their increasing convergence and integration, will have profound effects on corporate technology strategy and R&D organisation: We predict the necessity for firms cumulatively to acquire a widening set of in-house multidisciplinary capabilities and core technical competencies. This must be the case so that firms are able to understand, assimilate, fuse, and apply the new scientific and engineering insights and techniques into components, devices, systems, products, and services that cut across traditional industry boundaries.

Hypothesis 2: The identification, protection, and building-up of a continuously reassessed cluster of in-house scientific and technical competencies will necessarily proceed hand in hand with the formation of complex strategic alliances and networks with other firms, government research institutions, and universities internationally. This is so in order for firms to be able to access frontier scientific knowledge (especially the tacit aspects) and complementary technologies further downstream, in order to obtain adequate and timely return on investment in the market, and to facilitate interorganisational learning and knowledge creation and transfer.

Hypothesis 3: Given, first, the rapid and global nature of advances in the science-based generic technologies (in the same or other, related or converging fields); second, the path-dependent and historically accumulated core strengths of nations in scientific disciplines, technological trajectories and engineering skills (e.g., in miniaturisation, micro- and nanofabrication techniques, instrumentation and measuring techniques); and third, the scarcity of leading-edge scientists and associated world-class research teams and facilities able to conduct leading-edge research in specific, or groups of related, fields, we predict that transnational corporations (TNCs) will have no choice but to conduct increasing amounts of basic research outside their home economy, despite the historical importance of central corporate laboratories at home for the conduct of fundamental and strategic research. Increasingly TNCs will need to locate R&D facilities abroad to form deep linkages with such geographically dispersed sources, recruit local scientists and researchers, and participate in local science, academic, and industry networks embedded within different NSIs, while integrating and leveraging information and knowledge generated into their global R&D and corporate strategies. Local affiliates will have to be given more autonomy in the generation of new knowledge in foreign cultural and institutional settings. The integration of R&D across different cultures will present new management difficulties.

Hypothesis 4: National systems of innovation that possess world-class science and technology infrastructure and institutions engaged in frontier research will, first, play an increasingly important role in contributing sources of competitive advantage to domestic agglomerations of industries and firms, and, second, act as a magnet attracting inward flows of foreign direct investment and R&D facilities, which, together with parallel international alliances of domestic institutions, would enhance the generation of research externalities, local spillover effects, and economy-wide increasing returns for networks of interrelated firms and industries in the domestic economy, thereby reinforcing their role in competitive advantage.

Hypothesis 5: Despite the globalisation of technology and the implementation of global terrestrial and satellite digital communication networks, governments will acquire increasing importance and new roles in the ability of an NSI to create, maintain, or expand a national presence in specific technologies and industries. In particular, we ascribe an even greater role for governments in supporting the conduct of (increasingly) multidisciplinary basic research in local centres of excellence, in education and training programs, in the creation of a world-class national science and technology infrastructure (including communications and measurement and testing instrumentation), and in the diffusion and application of new research information and knowledge to domestic and local networks of industries, including services.

Spatial Agglomerations and the Location of Inventive and Innovation Activities in the Twenty-first Century

The key issue in international political economy today involves the question of where economic and industrial activity will locate in the future, under conditions of rising globalisation of scientific, technological, financial, and economic activities fuelled by a process of trade liberalisation and new transport and communications technologies. Transnational corporations play a crucial role in this process of ever closer integration of national economies, now moving beyond traditional trade-based linkages to more complex and deeper production linkages. Using advances in information and communications technologies, TNCs are beginning to employ integrated production strategies accompanied by a sophisticated intrafirm division of labour in corporate functions (e.g., production and R&D) and intricate network structures that are leading to new patterns in the global location of economic activities and that raise a new host of policy issues for national regulation, legal and fiscal frameworks, competition policies, and the protection of intellectual property rights. Indeed, is there any role for government intervention such that would attract, maintain, and expand the location of inventive, innovative, and economic activity in specific localities, domestic economies, and regions today and in the future?

A set of recent papers provides important insights on the interplay between globalisation and industrial location. Venables (1998) suggests that traditional explanations of changing patterns of trade and location based on the classical theory of comparative advantage are inadequate to the task, and need to be enriched by explanations from the side of the new economic geography and theories of cumulative causation. Audretsch (1998) provides an explanation for the observed importance of local proximity and geographic clusters, even in the presence of globalisation and TNCs as dominant forces in economic activity. The twin processes of globalisation and the telecommunications revolution (leading to the declining cost of information with distance) are, he argues, leading to a shift of comparative advantage in developed countries toward greater importance of knowledge-based innovative activities. However, knowledge is more efficiently transmitted via local proximity, while spillover effects cease as knowledge moves across geographic space. Hence, innovative activity based on new knowledge in high-technology industries tends to cluster in regional agglomerations. This is leading to new government policies at the regional and local level in order to facilitate the formation and the innovative performance of regional clusters (for evidence on six regional clusters in optoelectronics, and the extent to which they are embedded in national or international networks).

However, a basic proposition of this chapter is that in order to provide a better conceptualisation of the determinants of the agglomeration and location of economic and innovative activity, in the cases of both TNCs and high-technology entrepreneurship, and to devise policy implications for the role of the government in different NSI, we need to examine in more depth the nature of the scientific and technological revolutions under way. Our analysis is therefore a necessary complement to the literature on the new economic geography, cumulative causation, and national attractiveness and global competitiveness. This chapter is a modest contribution to these debates. We first examine some of the key characteristics of the three generic science-based technological revolutions, in the context of the hypotheses advanced above. We then ask what is the new learning on, and what are the modes of R&D organisation that enable companies to link technology to corporate strategy in a global context. We illustrate the discussion with the British firm Imperial Chemical Industries (ICI) research and technology strategy in a decentralised R&D structure, and the process whereby it manages its knowledge base in order to meet corporate objectives in global business. A second case study refers to the German firm Daimler-Benz's centralised R&D structure, which emphasizes the buildup of corporate-wide competencies in materials technologies over the long run. The third case study examines the U.S. firm Alcoa's buildup of core materials science and technology capabilities that have led to a successful alliance with the German firm Audi in lightweight car body structures. The latter case study breaks new ground in pointing to the future directions in the automotive industry and its relations with the supplier system globally, based on advances in materials technologies.

However, the strategies in the case studies, which due to space limitations focus on materials producing or using TNCs, have to be developed and implemented in the context of knowledge-based economies within different national systems of innovation; these are examined in the last two sections of the chapter.

Emerging Technologies

Information Technologies, Computers, and Communications

Recent years have seen the convergence of the previously separate areas of computers (hardware, software, services), communications (fixed wire-line, optical and radar, mobile, satellite, and computer and mobile networking) and media and content technology. These very fast developments are pushed forward by the continuous advances in the technologies underpinning these converging fields, namely, microelectronics, photonics, and radio. These rapid developments are accompanied by technological discontinuities in many areas, such as the arrival of interactive multimedia, a ceaseless effort to create corporate, local, and national networks and interconnections on a global scale and the arrival of low-cost portable computers, cellular phones, and satellite communications (in late 1998 and subsequently) that facilitate access to information, voice, fax, and text from almost anywhere, at any time over an emerging global telecommunications infrastructure. The value chains of the deliverables in this sector (software and electronic systems, packages of interactive multimedia) are complex. Underpinning the software and electronic systems and their distributed systems technologies are pervasive fundamental inputs (such as technologies of data compression, storage, sensors, display devices, radar, and power sources). These in turn depend on a multidisciplinary science base composed of computer science, mathematics, materials, microelectronics, photonics, control, instrumentation technologies, and communications. This convergence of previously separated fields and consequent discontinuities in digital communications networks, portability, and mobility and in media technology are poised to create conditions for a new phase of fast growth in these markets.

This sector is a global business sector dominated by large TNCs that would need to possess the financial, technological, and organisational capabilities to invest in future technologies, fuse complementary and converging technologies, and operate at the interface of previously separate industries and technologies. Critical issues arise here with regard to in-house skills and necessary global links and alliances (as evidenced by the formation of several global alliances by national telecom companies in recent years) for TNCs to remain competitive in their global business operations.

Equally, there are important issues surrounding the ability of NSI to provide the necessary conditions that will attract and maintain a strong local presence of key global players in these fields. The decision by Siemens of Germany and Fujitsu of Japan to close down two electronic factories in the United Kingdom in the 1998 illustrates the difficulties. We could conjecture that, in conditions of global difficulties faced by an industry, it could be that foreign production sites are easier to close down than those in the home economy, for political considerations.

The Biosciences Revolution

It is claimed that biology has emerged as the dominant science for the next century. Very powerful new research tools as well as deeper understanding of areas such as molecular and cell biology, genetics, and advanced instrumentation techniques are now making it possible to resolve biological questions (e.g., the functioning of the immune system) that would have been intractable even a few years ago. Moreover, the rapid advantages in information technologies and computing power and their application to, if not integration with, the life sciences, together with the completion of the deciphering of the human, plant, and animal genome, will lead to major developments in future applications of the life sciences. The rapid advances in genetic biological technologies are leading to great improvements in existing industries and to the creation of entirely new industries right across health care, medicine, food and nutrition, environmental management, chemical synthesis and manufacture, agriculture, and nonfood agricultural products.

For the purposes of this chapter, suffice it to point to the following:

1. The innovation process in health and life science industries involves strong links with basic research (possibly even stronger than other high-technology industries, including electronics). For example, basic medical research discoveries have a direct and immediate impact on diagnostics and therapeutics. This offers one part of the explanation of the observed tendency for new biotechnology and bioengineering companies to cluster in close proximity to academic centres of excellence in basic research, and, moreover, to maintain strong links with university departments and centres. New skills and competencies are, in addition, acquired by small startups and large companies through the transfer of key academic researchers from universities.
2. There is strong evidence that the links between the domestic science base and the innovation capabilities of domestic pharmaceutical and biotechnology companies are of paramount importance. It can be argued that it is the very strength of the United Kingdom's NSI science base (e.g., universities, medical schools' and medical charities' R&D, and their networks and linkages with the higher education infrastructure) that underpins a strong performance by the UK pharmaceutical and biotechnology sectors.

3. Both U.K. industry officials and the 1995 U.K. Technology Foresight Programme findings stress the need for the government to offer multidisciplinary support for integrative biology (linking gene function identification right through to a fuller understanding of neurobiology and immunology) and the vital importance of a modern university infrastructure and advanced computer skills. It is also stressed that the future strength of U.K. pharmaceuticals and biotechnology depends on educational programs that can deliver the human capital required in the cross-disciplinary fields underpinning the new technologies (e.g., bioinformatics) together with the creation of an efficient and up-to-date computer infrastructure for the conduct of research.
4. Biotechnology has moved from gene splicing and monoclone antibodies in the 1970s and 1980s to a rapidly advancing set of core tools and platform technologies, most of which did not exist even five years ago, with enormous implications for drug discovery and development, R&D organisation, and the formation of strategic alliances between large pharmaceutical groups and small biotech startups, and between them and academic and governmental organisations. These tools are predicated, among other factors, on genomics, combinatorial chemistry, high-throughput robotic screening, and bioinformatics. We can predict that all biotechnology and pharmaceutical companies must offer a strategic response to the developments in genomics as well as to the other revolutionary advances and complementary technologies in drug discovery and development. We can already identify several of the large TNCs in pharmaceuticals that have begun to incorporate the new developments into their in-house core competence acquisition, global R&D and location strategies, and strategic alliances with small biotechnology companies.

The Materials Revolution

The last decade or so has seen a revolution in the science and engineering base of materials design, processing, fabrication, and application. We can now begin to design materials on the basis of a fundamental understanding of their properties. This has important implications for the continuous, incremental improvement of existing, conventional materials up to their evolving theoretical limits and for the development of entirely new materials that go beyond such performance limits and are tailored to meet requirements in end use. In effect, like modern alchemists, material scientists and engineers can intervene and rearrange the atomic and molecular structure of matter, and control its form, properties, and performance in use. This process involves a close understanding of the relationship between a material's microstructure, the synthesis and processing path, the resulting properties and the material's performance in use, and its application across all materials classes: metals, polymers, ceramics, and composites of these.

Successful development of materials technology is a complex task involving the accumulation of in-house generic knowledge and skills in materials science and engineering disciplines, mathematical modelling and simulation and advanced testing, measurement, and characterisation capabilities, as well as the acquisition of core competences in specific materials, components, devices, and systems integration. Thus, it is likely that considerable investment in in-house knowledge acquisition and competence creation and development is essential for successful innovation and market differentiation in technology-based competition. However, alliances with universities, government laboratories, and other companies increasingly play an important role in knowledge creation and the core-competence acquisition agenda of firms. These developments lead to the proposition that firms and nations need to offer a long-term strategic response and to the prediction that strategic alliances, both vertical and horizontal, and global networks between materials producers and users will be an important feature of global technology strategies. Serious implications follow for the concepts of core competencies and competence-based competition. We have analysed this elsewhere.

R&D and Technology Strategy in a Global Context

Given the scientific and technological imperatives identified above, how should firms manage their R&D activities to achieve and sustain their competitive advantage? We provide first answers to this issue in this and the following sections through a literature review and case studies. Technology has become critical in creating and sustaining competitive advantage in the global market. The evidence, drawn from the accumulated experience of Arthur D. Little Inc. in this area, indicates that the most successful corporations are those that have developed a clearly defined technology strategy that is then fully integrated into corporate strategy and vigorously managed at senior-management level and, of course, communicated to all levels of the company. Most firms today, however, are unprepared for the R&D imperative they face. We can identify, in fact, three generations of R&D management. First, following World War II, the traditional and widely adopted approach was to treat R&D expenditures as overhead, with an annual budget set as a certain fraction of sales or some other performance indicator. The R&D unit enjoyed great autonomy in setting its research priorities, and there was no guarantee that these would conform to corporate or business objectives. Rising R&D costs, recession, and declining product life cycles have focused attention on evaluating the cost-effectiveness of the R&D investment expenditure. The move toward a more systematic approach in second-generation R&D led to the employment of "the net present value" role in order to offer financial justification for R&D projects. Although this development may make incremental contributions to value, it may have little long-term significance for the company. An additional problem is that, although each project may in itself be consistent with the objectives of the businesses, the total R&D portfolio may bear little or no relation to overall corporate strategy. And, while business managers and R&D managers certainly interact, the technology base of the firm as a whole is not clearly recognised or considered a strategic asset, and tends therefore to be eroded over time.

Individual project-based, quantitative, and more systematic second-generation R&D is still in widespread use today. However, according to Arthur D. Little, Inc., the intensification of technology-based competition requires a move to third-generation R&D management, where firms develop and integrate technology and R&D strategy with corporate strategy under strong senior management leadership. Such an efficient, strategically focused, and integrated R&D management can lead to incremental improvements in defending existing businesses or in generating technological breakthroughs that provide the basis for new businesses in the medium to long run.

Nevertheless, a number of writers have pointed out that it is insufficient simply to view technology strategy as the planning and management of technology in order to be consistent with and underpin the success of business objectives at the organisational level. Rather, corporate strategy itself, together with the associated general management practices, is a variable in the process of building and evaluating technology strategy. What distinguishes firms in their innovation capabilities is not only the efficiency by which they apply technology in the pursuit of business objectives but also the fact that corporate strategy itself will vary in its emphasis and ability to identify and utilise technological opportunities. That is, corporate strategy can be based on successful technology development and application, and provide the framework, organisation, and managerial practices for the strategic use of technology in order to achieve competitive advantage.

The process of strategy conceptualisation and implementation takes place, of course, within firms with already established organisational structures, management practices, and corporate culture. In this context, success requires an appropriate organisational structure, management process, and corporate culture as firms pursue a technology-based corporate strategy, in the sense that either technology is crucially important, or, more radically, it constitutes the very foundation of strategy. A key issue here is the role of the R&D function in the organisation whereby, on the one hand, it needs to specialise in a set of activities and, on the other, it has to be integrated with other functions.

The Strategic Management of Technology

The Need for a Global R&D Strategy

The complexity, multidisciplinarity, rapid pace, and global multipolar nature of scientific and technological advance require that firms form linkages with, gain access to, and participate in knowledge creation and distribution networks with other firms, academia, and government laboratories across different NSI. This necessitates the formation and implementation of global R&D and production location strategies that are strongly integrated with corporate strategic objectives at the senior-management level, and the management of in-house technical and organisational-learning competencies in complex alliances with other firms and international centres of research excellence. Thus, Toshiba's ability to develop next-generation semiconductor devices at the atomic scale depended not only on the accumulated knowledge base of its central laboratories in Japan and their networks within the Japanese NSI, but also on, first, the location of a basic research laboratory in Cambridge Science Park, in England; second, its alliance with renowned scientists in physics at Cambridge University; and, third, the networks formed by participating in frontier research within the U.K. national system of innovation. Moreover, the invention and subsequent commercialisation of the device depended on a string of global interfirm alliances in complementary technologies as well as the pooling of financial, production, and commercialisation costs and risks, as in previous generations of DRAM devices, but such approaches are now even more necessary.

Historically, firms have located development facilities close to manufacturing plants in order to assist the development and commercialisation of increasingly complex products. However, given the rapid advance of multidisciplinary scientific knowledge at several focal points internationally (in the same as well as related and complementary fields), TNCs have no choice but to form deep linkages with such sources, to engage actively in the process of information gathering, knowledge-creation, and learning, and to deploy these resources within corporate-wide networks and research programs aimed at meeting corporate strategic objectives. Here, the possession of in-house skills and competencies in attracting, forming, and managing formal and informal alliances in a process of continuous learning with other firms, universities, and government laboratories on an international terrain will be of utmost significance for the ability to conduct frontier research, to practice market differentiation, and to generate competitive advantage that can be sustained over time.

There is, therefore, an increasing need for companies to locate R&D facilities abroad, not simply to find out what research is being done (this does not necessitate location abroad today), but also to recruit local scientists and researchers and to form links with local science, university, and industry networks embedded in national systems of innovation. Companies need to participate in local networks and information flows and manage the two-way flows of tacit and codified knowledge, and to create new knowledge and integrate it into globally located R&D facilities and strategies. This does not necessarily conflict with the maintenance of central research facilities at the home base serving corporate needs as a whole, but there is an increasing need for greater decentralisation of basic research activities in terms of control and location, which may give subsidiary units more local autonomy while playing a full part in the design and implementation of corporate scientific knowledge creation and implementation strategy. Consequently, a firm's ability to access such networks and to combine or integrate interactive learning processes determines the economic influence of individuals and firms.

Levels of Strategy

Even if we broadly agree that a technology strategy is of critical importance for competitive advantage today, and that it must be fully integrated into corporate and business unit objectives, many difficulties remain. A number of writers have pointed out that, in recent years, firms have found it necessary to concentrate on the core (or strategic technology) assets they have accumulated over time and to use them for market differentiation and competitive advantage over time. On the other hand, there is also a noticeable tendency today toward the increasing complexity of products and processes that draw upon diverse technologies and their integration into systems. If we accept these two trends, there is an apparent conflict between in-house specialisation in core technology competences and skills and the need to continually broaden the in-house technology portfolio, as Jones, Green, and Coombs (1994) pointed out. This then gives rise to the necessity for companies to identify the technology assets of the firm and to manage them in a strategic context.

The management of technology in the context of business strategy would have different implications depending on whether we were dealing with the business units, the division, or the corporation as a whole. If, in order to concretise these ideas, we examine technology strategy at the business-unit level, then the R&D portfolio would lean heavily toward the business unit's market objectives and would, in the main, be financed directly from this business unit's operating budget. Technology strategy is thereby subjected to strong market pull forces. The focus on business-unit market position and financial returns could result in an overemphasis on short-term R&D objectives and on incremental process improvements at the expense of product development for the future.

However, if we examine technology strategy at the corporate level, then much would depend on the type of businesses owned or operated by the corporate parent, synergies across its subordinate units, and the type of strategic management control exercised by the corporate centre. Where the corporate centre exercises strong strategic planning or control across business units and divisions, and synergies exist among the latter, then we expect technology matters to play a greater role in corporate strategy. Here, firms of this type would have, or would have had in the past, central R&D facilities that conduct research programs to meet corporate strategic aims. Central laboratories would be developing generic scientific and technology competences that apply across business units and divisions, while also generating new knowledge, perhaps in collaboration with firms, universities, and government laboratories, which would create new business opportunities. Corporate laboratories may or may not be directly involved in near-market R&D and product development addressing the needs of specific business units.

Evidence collected about U.K. companies by Coombs et al. (1996) shows that, in the last two decades, firms with strategic corporate control and strong synergies in technology across business units have shifted toward greater decentralisation of methods of R&D funding and control. This can be seen in the greater emphasis placed on (1) R&D funding away from corporate to business unit sources, even if central labs have been retained, and (2) the growth of decentralised R&D labs and activities attached to divisional or business units. Again, this approach is very much in accordance with determined efforts by firms in recent years to make their R&D programs more market driven and commercially effective. A consequence of this drive is that central laboratories have been either run down or eliminated altogether. The process of decentralisation of R&D, however, and a focus on the business unit's short-run commercial objectives can leave such units vulnerable to the emergence of new technologies that pose a serious competitive threat to which a specific unit may be unable to respond by itself or by drawing on technologies from other units or a run-down or closed-down corporate lab.

These considerations point to a need for firms to exercise strategic control over R&D programs. One important response to this challenge has been the setting up of a corporate unit for the strategic management of technology. Such units are charged with the responsibility of continually examining the R&D portfolio, identifying/coordinating/disseminating the technology competencies and skills between business units, identifying and strengthening core technologies that are required across more than one business, integrating technology strategy with corporate strategic management objectives, and examining emerging technologies for their potential impact on existing or new businesses.

The next section briefly illustrates one such approach taken by ICI. This is counterpoised with the approach of Daimler-Benz within a centralised corporate R&D structure. The case studies on Alcoa and the Alcoa-Audi long-run R&D collaboration indicate the rising importance of building core technical competencies and getting close to your customers globally, organisationally as well as in terms of physical proximity to produce and deliver advanced Aluminium structures for Audi cars produced in Germany.

Chapter 7

Ascorbate Depletion: Step in Nickel Carcinogenesis

Nickel compounds are known to cause respiratory cancer in humans and induce tumours in experimental animals. The underlying molecular mechanisms may involve genotoxic effects; however, the data from different research groups are not easy to reconcile. Here, we challenge the common premise that direct genotoxic effects are central to nickel carcinogenesis and probably to that of other metals. Instead, we propose that it is formation of metal complexes with proteins and other molecules that changes cellular homeostasis and provides conditions for selection of cells with transformed phenotype. This is concordant with the major requirement for nickel carcinogenicity, which is prolonged action on the target tissue. If DNA is not the main nickel target, is there another unique molecule that can be attacked with carcinogenic consequences? Our recent observations indicate that ascorbate may be such a molecule. Nickel depletes intracellular ascorbate, which leads to the inhibition of cellular hydroxylases, manifested by the loss of hypoxia-inducible factor (HIF)-1[α] and -2[α] hydroxylation and hypoxia-like stress. Proline hydroxylation is crucial for collagen and extracellular matrix assembly as well as for assembly of other protein molecules that have collagen-like domains, including surfactants and complement. Thus, the depletion of ascorbate by chronic exposure to nickel could be deleterious for lung cells and may lead to lung cancer.

The high consumption of metal-containing products by industrial societies inevitably leads to environmental pollution by heavy metals, including nickel(II), chromium(VI), and cobalt(II), at all stages of production, use, and disposal. Human occupational exposure to metals occurs primarily via inhalation and ingestion of metal-containing dusts and is particularly high among stainless steel welders, miners, and metallurgy workers. The environmental exposure mostly results from burning of fossil fuels. Accumulation of metals in the human body can cause different diseases, including cancer. There is sufficient evidence in humans for the carcinogenicity of Cr(VI) and Ni(II) [International Agency for Research on Cancer (IARC) 1990], which are not known to be essential in humans, and limited experimental animal data support Co(II) carcinogenicity.

How can metals be carcinogenic? One possible answer is based on the ability of metals to generate reactive oxygen species and other reactive intermediates or react directly with DNA. Indeed, several transition metals can generate reactive oxygen species in biologic fluids at physiologic pH, but this does not explain why the relatively weak redox-active Ni(II) is as good a carcinogen as the oxidatively very active Cr(VI). Moreover, Cr(VI) is a better carcinogen than Cr(III), and yet Cr(III) is much better in DNA binding reactions than nickel or Cr(VI). In addition, the highly redox-active metals iron and copper, which also bind to DNA more avidly than does Ni(II), are only weakly carcinogenic, if at all.

Another explanation may be based on the known ability of carcinogenic metals to facilitate DNA damage through inhibition of DNA repair enzymes or binding to his tones. In both cases, it is assumed that the ultimate target of free radicals or metals is DNA and that the mechanisms of carcinogenesis must include genotoxic effects. However, metals such as Ni(II) are only weakly genotoxic/ mutagenic. In addition, one major requirement for Ni(II) carcinogenicity is prolonged action on the target tissue, which is typical of tumour promoters acting through epigenetic mechanisms, rather than tumour initiators, which are mutagenic. Therefore, Ni(II) compounds with limited solubility and long retention in biologic fluids are stronger carcinogens than are easily soluble compounds. So, considering the above, it is probably time to investigate how metals can promote carcinogenesis. Also, we should shift our attention from DNA to proteins and/or other molecules that offer a much broader and more vulnerable target for metal interactions. One such target is L-ascorbic acid, or vitamin C.

The notion that ascorbate is involved in resistance to neoplasms was introduced and advocated by Linus Pauling. This notion was based on demonstration of low ascorbate reserves in cancer patients, and later on results of clinical trials where survival time of cancer patients was prolonged by ascorbate supplementation. It is well known that transition metals can interact with and destroy ascorbate. The role of ascorbate as an antioxidant molecule has been appreciated for years, but only recently ascorbate was found to play a crucial role in hydroxylation reactions that determine interactions and functioning of thousands of cellular proteins. The hydroxylation is catalysed by a variety of nonheme iron-containing dioxygenases, including prolyl, asparaginyl, and lysyl hydroxylases, and novel DNA repairing enzymes, human ABH2 and ABH3 proteins, which require [α]-ketoglutarate (2OG) as cofactor. Most important, iron in these enzymes is maintained in the active Fe(II) form by ascorbate. Thus, the activity of these enzymes depends on two factors, iron and ascorbate. Both may be targets for Ni(II) and other carcinogenic metals. Can this lead to carcinogenesis? Here, we present some of the evidence suggesting this possibility. We discuss one possible way—there are certainly others.

"Oxygen Sensor" Senses Nickel

In rats, intrarenal injection of nickel subsulfide resulted first in a pronounced erythrocytosis and eventually in the development of local tumours. The high erythrocyte counts were mediated by increased erythropoietin (EPO) production. Exposure of human hepatoma cells (Hep3B or HepG2) to low oxygen atmosphere, or to increasing amounts of Ni(II) or Co(II), resulted in a dose-dependent enhancement of EPO mRNA expression. These findings led to the hypothesis that the induction of EPO by Co(II) or Ni(II) resulted from the substitution of the Fe(II) ion in an "oxygen sensor". Reduced oxygen delivery to the kidney is the physiologic stimulus to EPO production, and EPO transcription is now known to be under the control of the hypoxia-inducible factor 1 (HIF-1). HIF-1 DNA-binding activity is similarly induced by hypoxia or Co(II) chloride treatment, as originally described in Hep3B cells. In fact, exposure of cells to Co(II) was used to induce sufficient HIF-1[α] in cell nuclear extracts to allow biochemical purification.

The "oxygen sensor" hypothesis implies that Ni(II) may interfere with Fe(II) metabolism. Iron and nickel belong to the eighth group of transition metals, with atomic numbers 26 and 28, respectively. Fe(II) and Ni(II) have similar ionic radii of 0.74 [Angstrom] and 0.69 [Angstrom] and similar ligand affinity (Nieboer and Richardson 1980). This suggests that, in theory, Ni(II) could be accommodated at an enzymatic protein Fe(II)-binding site, and substitution is possible.

However, because Ni(II) is not as redox active as Fe(II), the substitution would inactivate the enzyme. A good example is phthalate dioxygenase that can be poisoned by divalent metals. This enzyme contains two iron-binding sites. A Rieske-type [2Fe-2S] cluster serves as an electron-transferring cofactor, and a mononuclear iron site is the putative site for substrate oxygenation, similar to that predicted in prolyl hydroxylases [prolyl hydroxylase domain (PHD) enzymes]. Any of the metal ions, Fe(II), Co(II), manganese(II), Ni(II), or zinc(II), can occupy the mononuclear site, but only Fe(II) is competent for effecting hydroxylation.

Although the substitution of metals in the enzymes seems possible, no experimental data indicate that this actually occurs in living cells. Most existing data were obtained in vitro using recombinant or biochemically isolated proteins. In living cells, the concentration of free transition metals is maintained at very low levels. This must be true for both essential and nonessential metals. Because most transition metals are dangerous for cells, they exist inside the cells bound to storage proteins or are transferred to the enzymes with the help of special metallochaperones. Such a model has been developed for copper, and undoubtedly similar mechanisms exist for other essential metals. Thus, at present, the hypothesis of in situ Fe(II) substitution by Ni(II) in the cellular "oxygen sensor" remains to be tested. However, besides the iron substitution, the poisoning of "oxygen sensor" may be explained in an alternative way when the critical role of ascorbate as an iron reductant is considered.

Depletion of intracellular ascorbate produces a phenotype observed in hypoxic cells or in cells with mutated von Hippel-Lindau (VHL) protein. Importantly, hypoxia is a common feature of neoplastic tumours. The molecular response to hypoxia is mediated by the HIF-1 transcription factor that stimulates angiogenesis, changes energy metabolism and iron homeostasis, and promotes cell survival in the affected tissues. Therefore, in order to survive, tumour cells should have high HIF activity. Like oxygen deprivation, exposure of cells to Co(II) or Ni(II) also results in the induction of HIF-1 and production of hypoxia-like responses.

HIF-1 transcription factor is composed of one [α] and one [β] subunit. The [α] subunit is the regulatory component of the HIF-1 complex and is unique to the hypoxic response. Under normoxic conditions, this protein is virtually undetectable in most cells because of hydroxylation and rapid proteasomal destruction, but it can accumulate after exposure to proteasomal inhibitors such as lactacystin or MG-132. Accumulation of the HIF [α] subunit in the presence of hypoxia or Ni(II) implies that proteasomal degradation of the protein is impaired by these exposures. The [β] subunit of HIF-1 is constitutively expressed. In hypoxia or after metal exposure, the [α] subunit dimerises with a [β] subunit and translocates to the nucleus. The produced HIF-1 complex binds to the HIF response element's initiating transcription of hypoxia-inducible genes.

The [β] subunit is also involved in xenobiotic responses where HIF-1[β] forms a dimer with the aryl hydrocarbon receptor (AhR). Therefore, an alternative name for HIF-1[β] is ARNT (AhR nuclear translocator). It is noteworthy that although ARNT expression itself is not directly affected by Ni(II), the expression of AhR-dependent genes can be significantly suppressed by Ni(II) exposure. The down-regulation of AhR-dependent genes has an important toxicologic implication. It may lead to a decrease in toxicant removal. Because cross-talk between the AhR-dependent pathway and HIF-dependent pathway has been described, it is conceivable that Ni(II) affects a major regulatory factor upstream of both pathways. Indeed, the similarity in effects of Ni(II), hypoxia, and the inhibitor of 2OG-dependent dioxygenases (dimethyloxalytglycine) suggests that iron and 2OG-dependent dioxygenases are involved in regulation of both pathways.

Hydroxylation-Dependent Protein-Protein Interactions

It has been shown recently that the key protein-protein interactions that govern the HIF-1 activation are regulated by oxygen through enzymatic hydroxylation of specific proline and asparagine residues. Hydroxylation of proline residues 402 and 564 in the oxygen-dependent degradation domain of HIF-1[α] leads to its interaction with the VHL tumour suppressor protein, a part of the ubiquitin-ligase complex. This is followed by ubiquitylation and rapid proteosomal degradation of HIF-1[α]. The structural basis for this specific interaction is provided by the introduction of hydroxyl group at the proline 4-position that then facilitates hydrogen bonding with VHL's Ser-111 and His-115 residues. This is sufficient to enable discrimination between nonhydroxylated and hydroxylated HIF-[α] (HIF-1[α] and -2[α]) proteins.

Besides proline, hydroxylation of asparagine and lysine residues was found to be important for protein-protein interactions. Thus, asparagine hydroxylation allows for complex formation between HIF-1[α] and the CREB-binding protein and p300 transcriptional coactivators that in this way also become involved in the transcriptional control of HIF-dependent responses. The hydroxylation of proline and lysine residues in collagen molecules serves to form and stabilise the collagen triple helices under physiologic conditions. Collagen chains that do not contain 4-hydroxyproline cannot fold into triple helical molecules that are stable at body temperature. Ascorbate plays a key role in the trimerisation step of three [α]-chains and/or in the subsequent triple-helix formation of collagen. The collagen superfamily today includes at least 27 collagen types with at least 42 distinct polypeptide chains, and > 20 additional proteins with collagen-like domains. Altogether, they make up 30% of cellular protein mass. In addition to collagens, proline hydroxylation probably affects interaction of many more cellular proteins, because 4-hydroxyproline residues are ubiquitous in proteins with collagen-like domains. Among such proteins are lung surfactant proteins A and D, also known as collectins, or collagen-like domain and calcium-dependent lectin domain proteins. Surfactant proteins A and D are found primarily in the alveolar fluid in lungs of mammalians. Their primary structure is highly conserved among different mammalian species. Similarly to the loss of hydroxylation in collagen, the failure of surfactant protein hydroxylation results in a decrease in melting temperature of 9°C and failure of assembly of the collagen-like domain.

In addition to surfactants A and D, C1q protein, the first component of complement, also known as defence collagen because of its involvement in the immunomodulation of inflammatory and allergic responses of the lung, contains 4-hydroxyproline in its collagen domain. Therefore, the loss of hydroxylation of C1q protein should affect inflammatory responses in the lung. At the tissue level, the loss of hydroxyprolines in the collagen-like domains of surfactants A and D could impair their functions. Surfactant inactivation and deficiencies have, indeed, been associated with a variety of lung diseases, including pneumonia, asthma, and acute respiratory distress syndrome. A similar pattern of lung diseases has been reported for workers occupationally exposed to Ni(II). Numerous data indicate that ascorbate deficiency may be an underlying factor in the pathophysiology of asthma and acute respiratory distress syndrome. They are in agreement with the hypothesis that Ni(II) may target ascorbate in the lungs of exposed individuals.

There are at least three functional classes of hydroxylases belonging to nonheme dioxygenases. Among them are two long-known collagen prolyl-4-hydroxylases, the more recently identified FIH-1 (factor inhibiting HIF), and PHD1-3, asparaginyl and prolyl hydroxylases, responsible for HIF-[α] protein hydroxylation. The third class, characterised most recently, includes alkyl DNA dioxygenases, such as the Escherichia coli AlkB dioxygenase and its two human homologues, ABH2 and ABH3, which facilitate a novel mechanism of DNA repair. In the presence of oxygen, these enzymes can specifically hydroxylate alkyl groups on 1-methyladenine and 3-methylcytosine. The requirement of iron for the reaction as well as inhibition of hydroxylase activity in crude cell extracts by iron chelators suggests that these enzymes are iron dependent. Inhibition by N-oxalylamino acid derivatives, first developed for use with collagen prolyl 4-hydroxylases, further implies that these hydroxylases are also dependent upon 2OG as a cosubstrate. The source of the hydroxyl oxygen is always (> 95%) molecular oxygen. In the course of the hydroxylation reaction, the enzyme-bound Fe(II) splits dioxygen into two atoms, being converted itself to Fe(III) or Fe(IV). The exact mechanism of the reaction is not known, but it is likely that after full cycle iron is returned to Fe(II). Any interference with the reaction, not clearly defined yet, will leave iron at a higher oxidation state. The role of ascorbate is to reduce iron back to Fe(II) and thus reactivate the enzyme. This scenario implies that iron depletion or oxidation leads to the loss of hydroxylase enzymatic activity and, again, points to ascorbic acid as a crucial player.

Extracellular Matrix, Ascorbate, and Tumour Promoters

The extracellular matrix (ECM) is composed of a great variety of molecules and includes collagen family, elastic fibres, glycosoaminoglycans and proteoglycans, and adhesive glycoproteins. ECM functions to bind cells together, to provide conditions for growth factor-receptor interaction, and to control growth and differentiation. It plays an important role in maintaining structural integrity and regulating cell polarity, migration, and survival. The major component of the ECM is collagen. In cutaneous tissues (i.e., skin), type I collagen accounts for 80-90% of all collagenous proteins in ECM. Defects in ECM can lead to developmental abnormalities and cancer.

Neoplastic progression is associated with specific changes in the ECM. These changes can be observed in very early stages of tumour development and involve, first of all, disorganisation of the basement membrane and a decrease in the deposition of a new collagen. The exposure to tumour-promoting agents is characterised by similar changes. For example, skin treatment with the tumour-promoting phorbol ester 12-O-tetradecanoylphorbol-13-acetate (TPA) leads to a reduction of the collagen content of the mouse dermis. Both collagen degradation and synthesis are affected by the carcinogens, but the effect on degradation is more pronounced. Because collagen assembly greatly depends on ascorbate, the decrease in the ascorbate level during spontaneous or carcinogen-induced tumour progression could be a key factor in the ECM disorganisation. Indeed, it was shown that TPA decreases transport activity of both sodium-dependent vitamin C transporters SVCT1 and SVCT2. TPA-induced depletion of intracellular ascorbate can be surmounted by the addition of ascorbate that is manifested by the reversion of the cell-transforming activity of TPA.

The reduced collagen expression at the early step of carcinogenesis may cause disturbed keratinocyte adhesion to the basement membrane. In in vitro experiments, anchorage-independent growth represents a good test for malignant transformation of cells, emphasizing an important role of cell attachment and ECM in the carcinogenesis process. Indeed, the transformed phenotype of mouse fibroblasts that did not grow in semisolid medium could be completely reverted by ascorbate exposure. Moreover, addition of ascorbate inhibited soft agar growth of rat osteosarcoma or human neuroblastoma cells.

Maintenance of Ascorbate Levels

Ascorbate is a well-known antioxidant required by all mammalian cells for proper functioning. Although rodents (except for guinea pigs) produce ascorbate in the liver, humans are unable to synthesize ascorbate. This inability is likely to be reflected in the ascorbate levels in the blood serum, given normal consumption of ascorbate-containing food. The comparison of ascorbate levels in different species suggests that humans probably have lower levels than do rodents. Thus, a number of studies have provided values of 14.9-58.8 [micro]M ascorbate in blood serum for healthy volunteers receiving the recommended dietary allowance of 60 mg vitamin C daily. In another study, Dawson et al. (1999) described average levels of ascorbate of 35-40 [micro]M. In rats, Gumuslu et al. (2000) reported an average ascorbate level of 63 [micro]M in the blood serum, and in mice, Atanasova et al. (2004) and Kuo et al. (2004) found between 96 and 125 [micro]M ascorbate.

These species-related differences in ascorbate levels should be taken into consideration when results of animal experiments are analysed. For example, the inhalation exposure to carcinogenic nickel compounds resulted in increased incidence of alveolar/bronchiolar neoplasms and adrenal medulla neoplasms in male and female rats but not in mice; unfortunately, in this study ascorbate levels were not analysed in blood or tissues of experimental animals. Under our concept, it is conceivable that a higher level of ascorbate in mice protected these animals from nickel-induced tumour development. Additionally, the appearance of adrenal medulla neoplasms in exposed rats can also be mediated by the effect of Ni(II) on ascorbate levels because adrenal medulla has the highest requirement for ascorbate in the body. Exposure to cobalt sulfate causes alveolar/bronchiolar neoplasms both in rats and in mice, and cobalt sulfate is more efficient in the depletion of intracellular ascorbate. Could it contribute to the effectiveness of cobalt in animal carcinogenesis?

The maintenance of necessary intracellular ascorbate level in humans depends on dietary intake and subsequent distribution into cells. Thus, the intracellular level of ascorbate strongly depends on the ascorbate levels in the blood or in the culture media. Ascorbic acid and its oxidised form, dehydroascorbic acid (DHA), both can be transported into and out of cells. Glucose transporters GLUT 1 and GLUT 3 mediate transport of DHA, whereas ascorbic acid enters cells through sodium-dependent transporters SVCT1 and SVCT2. Inside the cell, DHA is immediately reduced to ascorbic acid, which accumulates at higher than outside concentrations.

The intracellular level of ascorbate is therefore regulated by its oxidation status, uptake, and efflux. In culture medium, the only source of ascorbate is serum. Thus, we found the concentration of ascorbate in Eagle minimal essential medium supplemented with 10% serum to be between 1 and 5 [micro]M. At the same time, human lung epithelial cells grown in this medium accumulated up to 60 [micro]M ascorbate. Nonetheless, it is likely that cells in tissue culture do not receive adequate amounts of ascorbate. Therefore, the ascorbate values reported for cells in vitro probably are lower than those existing in cells of corresponding tissues in viva. This may explain rather high [K_m] values reported for prolyl hydroxylases. Thus, the [K_m] values found for PHDs were found in a range of 140-180 pM, whereas for collagen prolyl-4-hydroxylase [K_m] was 300 [micro]M. The activity of DNA dioxygenases ABH2 and ABH3 was also found to be ascorbate dependent; however, the [K_m] values for the enzymes were not measured. The maintenance of intracellular level of ascorbate therefore is critical for the activity of specific hydroxylases.

Ascorbic Acid Interactions with Transition Metals

The intra- and extracellular levels of ascorbate can be greatly affected by the interaction with metals. In the absence of molecular oxygen, the reactions of ascorbic acid with certain transition metal ions, especially Fe(III), Cu(II), mercury(II) [Hg(II)], or Cr(VI), result in their reduction and oxidation of ascorbic acid to DHA, for example, 2Fe(III) + ascorbic acid $\rightarrow$ 2Fe(II) + DHA + [$2H^+$]. However, in the presence of oxygen or other oxidants, such as hydrogen peroxide or lipid peroxides, these metals act predominantly as catalysts in a chain of partially radical reactions. The metal catalyst cycles between its original and lower valency state. These reactions include the formation of a binary complex between the metal and mono-deprotonated ascorbic acid molecule, and dioxygen coordination by the bound metal. Once ascorbate is oxidised, the metal is dissociated because it cannot be bound to DHA.

Other metals, such as Zn(II), Ni(II), Co(II), or lead(II) [Pb(II)], which do not accept electrons at neutral pH, can neither oxidise ascorbic acid by themselves nor serve as catalysts for oxidation with molecular oxygen. However, this is only true for simple aqueous solutions in which only binary metal-ascorbate complexes can be formed. In tissue fluids and cell culture media, the situation is radically different owing to the presence of a wide variety of natural metal complexing agents (chelators), such as amino acids, peptides, proteins, nucleotides, and glutathione, which may bind and profoundly affect the redox potential of these metals or their dioxygen coordination capacity and thus make them catalytically active toward ascorbate oxidation. Most important, such ligands may form ternary complexes with the metal and ascorbic acid. Among natural chelators, histidine and histidine-containing peptides, especially those having the N-terminal XaaYaaHis motif (e.g., albumin), play the major role in biologic binding of transition metals such as Cu(II), Ni(II), or Co(II). The ascorbic acid in these complexes was found to be oxidised by ambient oxygen much faster than was ascorbic acid alone. Besides histidine, other amino acids, including glycine, alanine, serine, phenylalanine, and glutamic acid, have been found to form ternary complexes with Ni(II) and ascorbic acid, with stability of those with histidine and glutamic acid being highest.

Transition metal chelation by components of biologic fluids other than ascorbic acid may also facilitate the oxidative destruction of the latter through more indirect mechanisms, for example, those involving Fenton-like chemistry with metabolic [H_2][O_2] and generation of reactive oxygen and other radical species from lipid peroxides. These are avidly scavenged by the ascorbate anion, thus lowering its tissue levels. The best examples of this effect would include facilitation by Ni(II), a non-Fenton metal, reactivity with [O_2] or [H_2][O_2] through chelation with polyglycyl, polyalanyl, or histidine-containing oligopeptides.

In cells, the oxidation of ascorbate to DHA is readily reversible owing to specific enzymatic systems, including DHA reductase, glutaredoxins, and protein disulfide isomerase, and depends on glutathione or other thiols as electron donors. However, unlike ascorbic acid, DHA undergoes relatively fast ([$t_{1/2}$] ~ 2-15 min at pH 7) spontaneous hydrolysis to a product, 2,3-diketogulonic acid, which cannot be regenerated. In addition, DHA and 2,3-diketogulonic acid may also be further oxidised by molecular oxygen, [H_2][O_2], and some metal oxyanions (permanganate, chromate) to products such as threonic acid, oxalic acid, and various smaller products (e.g., carbon dioxide) that cannot be reverted back to ascorbic acid. There also is a possibility that metals can form complexes with derivatives of ascorbic acid, such as ascorbate-2-sulphate, 2-O-[β]-glucuronide, and 2-O-[α]-glucoside that have been found in mammalian cells, and ascorbate-6-phosphate, which has been found in bacteria but has not yet been detected in mammalian cells.

The Key Role of Ascorbate in Nickel Toxicity and Carcinogenesis

The effect of Ni(II) compounds on cells is complex and most likely involves modulation of many metabolic pathways. Recently, we found that Ni(II) exposure greatly depleted intracellular ascorbate, which results in the activation of the HIF-1 transcription factor and up-regulation of hypoxia-inducible genes. The addition of ascorbate to the media caused an increase in intracellular ascorbate and reversed both metal-induced stabilisation of HIF-1[α] and HIF-1-dependent gene transcription. Recently, the important role of ascorbate in down-regulation of HIF-1[α] level in tumour cell lines has been demonstrated. The depletion of intracellular ascorbate, as the initial step of Ni(II) toxicity, has profound effects on cellular and tissue metabolism. It shifts cellular homeostasis toward a more transformed phenotype.

Here, we are considering three major modifications in cellular metabolism due to the depletion of intracellular ascorbate by Ni(II): a) induction of hypoxia-like stress, b) impairment of assembly of proteins with collagen-like domains, and c) inactivation of ascorbate-dependent DNA repair enzymes. Exposure of human or rodent calls to Ni(II) stabilises HIF-1[α] and induces HIF-1 transcriptional activity, followed by the induction of glycolytic enzymes and glucose transporters. The up-regulation of the glycolytic pathway may be beneficial for cell survival, and if Ni(II) exposure is prolonged, it may select cells that maintain a high glycolytic rate and thereby acquire a phenotype similar to cancer cells, as described by Warburg (1956). In addition, other hypoxia-inducible genes, including BCL-2 binding protein Nip3, EGLN1, hypoxia-inducible gene 1 (HIG1), prolyl-4 hydroxylase, and cyclin G2, and growth factors, such as platelet-derived growth factor BB and vascular endothelial growth factor (VEGF), are up-regulated by Ni(II) in an HIF-dependent manner.

Although the benefit of up-regulation of some hypoxia-inducible genes for cell survival and transformation is not understood, the upregulation of others is clearly beneficial. Thus, up-regulation of cyclin G2 ceases cell proliferation, allowing conservation of energy. The up-regulation of growth factors has an antiapoptotic effect and helps to sustain cell growth by both autocrine and paracrine regulatory mechanisms. In the long term, VEGF as a major tumour angiogenesis and vascular permeability factor can support cell survival and growth via enhanced blood supply. Angiogenesis can be further stimulated by the suppression of antiangiogenic factors such as thrombospondin I. Both Ni(II) and hypoxia suppress thrombospondin I expression. The importance of HIF activation and expression of HIF target genes for evolution of a more malignant phenotype was shown in an animal model system. Additionally, immunostaining studies using monoclonal antibodies against the HIF [α]-subunits demonstrated increased HIF-1[α] expression in about 53% of malignant tumours, including colon, breast, gastric, lung, ovarian, pancreatic, prostate, and renal cell carcinomas, melanomas, and glioblastomas as compared with the respective normal tissues. In another study, increased protein levels of both HIF-1[α] and HIF-2[α] were detected in about 54% of tumours screened.

Similarly to hypoxia, Ni(II) exposure upregulates HIF-1 and transactivates HIF-1-dependent genes that are critical for invasion and metastasis that can be facilitated by the impairment of ECM. It is expected that collagen assembly will be affected even before HIF activation because the [K_m] for ascorbate of collagen hydroxylase is twice as high as the [K_m] for HIF hydroxylases. Nothing, however, is known about the effect of Ni(II) on collagen hydroxylation and assembly. The destruction of ECM selects for cells that can proliferate without attachment, which can be tested as the loss of anchorage dependence. Indeed, the loss of anchorage dependence is well documented in human and hamster nickel-transformed cells.

Almost all nickel-transformed clones are also capable of producing tumours in nude mice. Recently, we have shown that the expression of HIF-1[α] is required for the induction of soft agar growth by Ni(II). Under the concept of our hypothesis, it is conceivable that the disruption of ECM and activation of hypoxia-like stress inside cells will provide perfect conditions for selection of cells with neoplastic phenotype. Additionally, inactivation of DNA dioxygenases may introduce mutations preserving selected phenotypes. Nickel is a weak mutagen; therefore, one assumption would be that the [K_m] for ascorbate for these enzymes is rather low. Alternatively, DNA dioxygenases repair N1 alkylated adenines, which are not major mutagenic lesions.

If the hypothesis discussed here is true, one may predict that supplementation with ascorbate should alleviate carcinogenic effects of Ni(II). Although the direct data are not available and further investigations are needed to demonstrate protective effect of ascorbate in Ni(II) carcinogenesis, many data demonstrate preventive effects of ascorbate supplementation against Ni(II) toxicity: prevention of oral Ni(II) toxicity in weanling rats; suppression of nickel-induced lipid peroxidation in human platelets and placenta and in mouse and rat liver; decrease in nickel-induced cytokine production and elevated serum levels of hepatic transaminases in rats; reversion of nickel-caused decrease in hepatic glutathione level and superoxide dismutase, catalase, and glutathione peroxidase activities in rats; alleviation of clastogenic effects of Ni(II) in murine bone marrow; and prevention of DNA damage and increase of viability of human lymphocytes cultured with Ni(II). Osipova et al. (1998) proposed using ascorbate as an antimutagenic agent on nickel-exposed workers. It is noteworthy that in contrast to reports of aggravation by ascorbate of nickel-mediated in vitro toxicity, consistent with the known generation of reactive oxygen species in metal-ascorbate solutions, no such effects have been thus far observed in nickel-exposed and ascorbate-supplemented cells and tissues.

Recent investigations clearly show that the cellular ascorbate is greatly depleted by exposure to Ni(II) compounds, which results in the inhibition of prolyl hydroxylase activity and activation of HIF-1[α] and expression of hypoxia-inducible genes. The addition of ascorbate reverses these as well as a variety of other toxic effects of Ni(II) in cultured cells and animals. All this indicates that ascorbate may be the major target in nickel-induced toxicity and carcinogenesis. However, the hypothesis of the key role of ascorbate in Ni(II) carcinogenesis awaits further testing because the critical experiments are yet to be performed. Among them, most important is testing nickel carcinogenicity in mice or rats with mutations in L-gulono-[c]-lactone oxidase that make them unable to synthesize ascorbate. Monitoring levels of ascorbate during cell transformation in vitro or during tumorigenesis in vivo is another important task. The elucidation of the role of ascorbate depletion in the development of nickel-induced neoplasms could serve as a basis for prevention and treatment of Ni(II) and, perhaps, also other metal-induced human cancer. Bearing this in mind, we quote Cameron et al. (1979), from a historical paper published 25 years ago: "We agree and believe it to be essential that extensive studies of ascorbic acid in cancer be made without delay."

Cleaner Supply Chain: the Case of Green Chemistry

There is a great deal of interest in the development and deployment of green technologies and the actions required on the part of industry, academia, governments and civil society to drive them forward. This chapter uses the case of green technology in the global chemical sector to better elucidate the challenges of implementation of innovations for sustainable development, to analyse which approaches have been effective, and to provide generalisable knowledge about the types of strategies required to move these technologies from niche applications into widespread use.

For green chemistry, and innovations for sustainable development more generally, there is a need for greater public intervention, including regulatory regimes that are strictly enforced, investment in basic research and education to build human capacity, more outreach programs in collaboration with industry to aid with technology transfer and implementation, and economic incentives for firms that may have the desire but not the financial capacity to make use of these innovations. Voluntary collaborations and the influence of major supply chain actors, on their own, are not powerful enough to catalyse the increases in scale that are needed for a real transition to sustainability.

The past century has been an era of technological miracles. Advances in transportation, communications, agriculture and health have drastically improved the quality of life for billions of people around the globe. Malthusian predictions of exploding populations pushing the earth's resources to their limit have yet to materialise, thanks in part to the genius and dedication of generations of scientists, engineers, industrialists, entrepreneurs and public servants. Still, many believe the planet is in a more precarious position today than ever before; population growth, while slowing, continues to put pressure on resources. Increased standards of living compound these stressors as industrial activity intensifies to meet the growing demand for goods and services. Many technologies are a so-called double-edged sword, contributing to an overall improvement in quality of life while simultaneously degrading public health, deteriorating the environment and threatening the sustainability of resources.

One of the foremost technological challenges confronting the world today is the ability to sustain improvements in quality of life without permanently compromising the human and natural systems on which we rely. This is the underlying goal of sustainable development. The transition to a more sustainable trajectory is complex and requires a variety of responses, but one critical element of a more sustainable future is that we continue to harness the power of innovation. While technological innovation is a source of many of the environmental challenges that we now face, it is an integral part of development. Technological advancement has not been without cost, but its benefits are undeniable. For example, it is unimaginable to envision returning to an era before modern medicine or high-yield agriculture. A return to pre-industrial society is, quite simply, no longer an option.

This means that our future technological trajectory has to not only increase benefits, but also reduce costs. In other words, we need innovation for sustainable development. Such innovations improve the quality of life through efficiency and performance improvements, while at the same time reducing negative impacts on humans and the environment. These innovations are more than just abatement technologies, like sulfur dioxide scrubbers or effluent treatment systems; environmental performance is inherent in their design via a systemic, life cycle approach to development, adoption and diffusion.

Overall, innovations for sustainable development require complex analysis, interdisciplinary thinking and greater engagement with the underlying social, political and natural systems in which they occur. One reason is that technological innovations are a global public good. A challenge of global public goods is that there is a mismatch between those who pay the bulk of development costs and those who reap the benefits. The costs are usually incurred locally and privately, but the benefits may be felt globally. Another reason is that innovations themselves can be complex and uncertain, as innovators take into account a whole set of technical, political and social considerations.

For new technologies, net costs and benefits are usually unclear. There are questions about the risks of unintended impacts on people, the environment and even other products in the market. Complex situations can arise where a social benefit might generate an increased economic cost, or where one environmental improvement requires environmental costs in another area, such as decreases in emissions requiring increases in energy use. Decisionmakers are often faced with difficult trade-offs.

There are many areas in which innovations for sustainable development have begun to emerge. In recent years, for instance, there has been a great deal of investment in energy technology. Important innovations have also occurred in transportation, materials, water purification and water use. These advances have all been important, but they have taken place, for the most part, in isolation. Taking the global production-consumption system into consideration, this approach is a potentially haphazard way of evaluating and implementing changes in areas where large and broad impacts are sought. For finished products, a major challenge is the complexity of manufacturing supply chains. For instance, while electric or hybrid vehicles may reduce the energy required for transportation, the vehicles themselves are still very intensive in terms of the resources required to produce them. This highlights the need to think about innovations for sustainable development throughout the entire life cycle of products.

There are many innovations that have the potential to generate important positive impacts. This chapter will explore one particular set of innovations for sustainable development in the chemical sector. These innovations, known as green chemistry, are important not just for improvements from the perspective of sustainability, but also because they are integral pieces of the supply chains that produce some of the most products necessary for our day-to-day lives.

Green technologies in the chemical sector have an impact on a range of areas from building materials, textiles and fuels to plastics, personal care products and cleaners. In other words, chemicals are at the heart of many of the products needed to improve the quality of life of people around the world in the face of increasingly constrained resources.

Green chemistry also gives a model for how rapidly developing countries, such as China and India, can follow a sustainable trajectory of growth. It provides a pathway for the continued expansion of industry and economic growth, and the ability to provide key goods and services without permanently damaging the environment or depleting natural resources. Ideally, green chemistry and similar technologies have the potential to help reduce the tension that exists between the need for economic growth and the recognition that such growth often comes at the expense of the environment, health and the provision of basic necessities, such as clean air or drinkable water.

This chapter will describe green chemistry in more depth, including some of the barriers to its implementation in the United States and China. It will then describe policy approaches that have been used to tackle these challenges. Finally, the chapter will draw lessons about what strategies will be required in the future to increase the impact of green chemistry and other innovations for sustainable development.

Green Chemistry: Sustainability through the Supply Chain

One sector in which the life cycle concept is being taken very seriously is chemicals. Chemical production is a major industry not just in economic terms, but also in terms of reach. In the United States, it is estimated that nearly 96 percent of the economy is impacted by the chemical sector in some fashion. From basic petrochemicals all the way through to pharmaceuticals, chemicals play a vital role in manufacturing key goods. They play a role in energy generation (biofuels and petroleum are both parts of the chemical industry), agriculture, transportation, clothing, building materials, furnishings, consumer health and hygiene products, paper goods, electronics, plastics—the list is nearly endless. Chemicals are a global industry with complex supply chains that stretch across borders. A large portion of new capital investment in the chemical industry is occurring in developing countries. Despite being, in some ways, an established and entrenched industrial sector, the chemical industry also retains a strong tradition of innovation, entrepreneurship and growth. As such, innovations for sustainable development in the chemical industry have the potential to serve as an important foundation for a system of production that is more sustainable overall.

Green chemistry is the science behind bringing sustainability into chemical products and processes. The Twelve Principles of Green Chemistry were first defined in the mid-1990s by Paul Anastas and John Warner. While many of their underlying scientific concepts had been the subject of research for years, or even decades, Anastas and Warner were among the first to look comprehensively at the chemical life cycle and to think step by step about how to make each part of the process more sustainable. Over the years, the concept has gained traction in scientific and industrial chemical circles, leading to both academic research and actual implementation and production based on the Twelve Principles framework.

Green chemistry, which can be thought of as the "design of chemical products and processes that reduce or eliminate the use and generation of hazardous substances," is a design framework. While one critique of so-called green innovations has been that they provide reduced performance at increased cost, the explicit goal of green chemistry is to improve performance, economics and environmental impacts. In other words, products or processes based on the Twelve Principles must be competitive in the market. This is especially true for chemicals, which are often intermediate inputs along the supply chain and invisible to the end consumer, making it difficult for firms to demand price premiums for greener products.

If a set of innovations can increase profit margins and provide environmental benefits, then it would seem logical that adoption of these innovations would be widespread. An important question, then, is why development and implementation of green chemistry has been slow despite the potential for private and public benefit. There are no definitive statistics on how much of the chemical industry's production is based on green chemistry, but the current best estimate is somewhere around 1 percent. This does not signal widespread penetration of sustainability concepts into the industry, despite fifteen years of work on the part of academics, regulators and industry to move the science out of the laboratory and into production. The importance of the chemical industry and the existence of a well-defined innovation design platform make green chemistry an important area for study. While the dynamics of innovation in this particular case are not completely generalisable and do not capture all of the challenges or strategies for successful innovation of global public goods, they do encapsulate some important lessons for innovators, regulators and decisionmakers interested in a range of technologies. Previous research has identified some of the major barriers to widespread implementation of green chemistry, as well as the kinds of policy solutions that have been attempted over the past fifteen years.

The major barriers to green chemistry vary to some extent from country to country, depending on the different contexts in which the technologies are being developed and implemented. In the United States, for example, the chemical industry's capital infrastructure is well-established and new investment can be difficult; there are definite problems of technological lock-in, wherein decisions made years ago limit the ability of firms to easily innovate. This is especially true in mature industries like chemicals. In China, conversely, there is less of a problem with legacy capital and technologies since the Chinese chemical industry is currently in a period of growth and investment and thus has greater flexibility to implement new innovations, especially in terms of process. China does, however, face barriers to implementing green chemistry stemming from competing government agendas, and needs to develop technical and managerial capacity in this sector. China's legacy challenges are the result of policies and bureaucratic incentives that have prioritized economic growth over environment and safety. Another challenge to overcome is that, until recently, the Chinese domestic chemical industry had almost no in-house research and development. As a result, there are few experienced scientists, engineers and managers available to firms in this area.

There is also variation in the focus of green chemistry innovations in different countries, which reflects underlying policy contexts. Based on interviews with Chinese green chemistry leaders from industry, academia and government, the focus in China is on increased economic and energy efficiency. This stems largely from tough energy efficiency targets in the current five-year plan, which aims to improve energy consumption per unit of GDP by 20 percent. The expression of this target in the five-year plan has led to more public and private investment in green chemistry innovations related to feedstocks and industrial production.

In the United States and Europe, there is greater concern for the long-term negative impacts that persistent, bioaccumulative and toxic chemicals can have on humans and the environment. Policies at the state, national and regional level are being developed and implemented to control risks from chronic exposures to low levels of hazardous chemicals. This is reflected in more work being done not just on greener production processes, but also on alternative products and compounds that reduce overall hazard during their use and disposal.

Policy Strategies to Advance Innovation for Sustainable Development

The range of barriers to development and implementation of green chemistry has resulted in a multitude of policy strategies. Green chemistry addresses a number of environmental issues, including climate change, clean air and water, toxic pollutant reduction and resource efficiency. Its inherent inter-disciplinarity, as compared to environmental policies that are often focused on single media like air and water, has resulted in a situation where policy drivers originate from various levels of environmental regulation. Policy is driven not just by involuntary, government-mandated rules and programs, but also through a variety of collaborative, voluntary efforts on the part of government, industry, academia and civil society. Ten approaches derived from the experience of green chemistry can be applied to other instances of innovation for sustainable development:

1. Standards, certification and labelling
2. Databases, tools and information sharing
3. Awards and recognition
4. Economic incentives
5. Market creation and preferential purchasing
6. Technology mandates
7. Regulatory restrictions
8. Research funding
9. Partnerships and collaborations
10. Technical assistance.

Standards, certification and labelling are a set of tools most commonly used to deal with problems that arise from information asymmetries, which generate market failures and/or missing markets. Consumer-facing labels and the underlying standards and certifications can increase customer awareness and signal otherwise hidden attributes of a product, such as environmental impact, to consumers. Other types of standards, certifications and labels are designed to relay information throughout the supply chain. For green chemistry, a rigorous third-party green chemical information standard is still undergoing development in the United States. In the absence of such a program, innovators can face challenges communicating the green chemistry-related benefits of a product to potential customers. There is also space for differing green chemistry claims, some of which may not be valid. (This is sometimes referred to as "green washing.") Green chemistry has many dimensions and is designed to encourage continuous improvement. Developing certification has been a complex and difficult process and it remains unclear what the overall impact will be in the future.

Access to databases, tools and information sharing networks is cited as a major challenge by many of those involved in green chemistry efforts. Development and implementation of green chemistry innovations hinge on the availability of scientific and technical information. There is a large body of public knowledge developed in academia, as well as information accumulated by regulatory agencies, that could be made readily and easily available to academic and industrial researchers. Some of this is available via journals or proprietary databases, but it can be difficult to locate all of the information needed to successfully design and implement a green chemical product or process. Within the United States, several states, including Michigan and California, are working on databases and information portals to centralise and streamline access to key information. There are also multi-state efforts underway with similar goals. In the European Union, chemicals are required to be registered under the Registration, Evaluation, Authorisation and Restriction of Chemicals (REACH), a regulation creating another important public information resource. In addition, industrial organisations such as the American Chemical Society Green Chemistry Institute's Pharmaceutical Roundtable, have worked to put together common tools that are shared between members and are useful for specific chemical sectors (e.g., pharmaceuticals.

Awards and recognition programs for green chemistry are both long-standing and growing in popularity. In the United States, one of the most well-established policies to promote green chemistry is the Presidential Green Chemistry Challenge, which dates back to 1995.

This is one of only a handful of awards given out by the President. In India, the University of Delhi gave out its first set of green chemistry awards to businesses and academics in January 2008, which was followed by another set of industrial awards by Newreka, Inc. at the 2009 Industrial Green Chemistry workshop in December 2009 in Mumbai. Winning awards represents an opportunity for positive press and increases awareness of green chemistry among competing organisations and within the larger scientific community. Savvy marketing departments can also use them to enhance the image of a firm. Moreover, awards engender competition between firms to develop and implement green chemistry innovations. Because they are relatively inexpensive to administer, awards are used strategically at the state level in the United States, most recently in Michigan.

Some governments have been willing to provide direct economic incentives to promote innovations in green chemistry. In August 2010, France announced that it would invest 1.35[euro] billion in five key technologies, one of which is green chemistry. This investment will largely be in the form of loans and about a third will be in the form of grants. The government's goal is to encourage work on technologies that might be high risk, but that also have the potential for high impact. Other countries have more dispersed economic incentive programs. In the United States, some government investment in green chemistry innovation comes from programs supporting particular technologies, such as economic incentives for a variety of alternative energy programs that intersect with green chemistry innovations. Specific green chemistry innovations are also eligible for government grants available through different agencies' Small Business Innovation Research (SBIR) programs. These types of initiatives go above and beyond traditional methods to encourage innovation more broadly, such as corporate tax credits for research and development. This background level of government investment in innovation can also impact the economic feasibility of green chemistry implementation by impacting the general costs of research and development. Unfortunately, because many of these economic incentives either come from different parts of the government (United States) or are very new (France), there is little to no data available about their overall impact on the commercialisation and penetration of supported technologies into the wider marketplace.

For some innovations in this space, especially finished goods and products for use by businesses and/or individuals, the market is often not large enough to justify investment on the part of producers. However, there are examples in which the creation of new markets had a dramatic impact on the success of a green chemical technology. One example is the impact that Wal-Mart had on the market for biodegradable, corn based poly-lactic acid (PLA) polymers in the United States. In 2005, Wal-Mart announced that the plastic "clam-shell" packaging for fruits and vegetables in its stores would have to come from renewable resources. PLA maker Nature Works, LLC (formerly Cargill-Dow) was the only large PLA manufacturing facility in the United States that was producing the polymer from a corn feedstock. Since 2005, the market for PLA worldwide has grown at a compounded annual growth rate of 20.1 percent, largely driven by increased demand from the packaging sector.

Governments, which are themselves large purchasers with significant market power, can use their procurement policies to create demand for innovative green chemistry products. When government purchasing guidelines create a preference for particular green chemistry products, they can, in some circumstances, provide enough evidence of a market to justify investment by firms in adopting specific technologies. The challenge has been for governments to identify the appropriate criteria for a green chemistry product. This is an example of a situation that may require multiple policies in the absence of standardisation. The problem would be eased by the existence of a green chemistry standard, certification and labelling system, which would make the task of establishing appropriate criteria much easier and more effective for government entities at all levels.

Technology mandates have for a long time been used in environmental regulation. This more stringent version of market creation, in which a government dictates to firms what specific green chemical technologies must be implemented, has several drawbacks. First of all, it is often expensive. Even more importantly, it is inflexible. By mandating a particular technology, firms are forced to invest financial and human resources into something that might not be the most cost-effective strategy for their situation. Technology mandates are especially difficult in the context of green chemistry, which by definition is amorphous and has a particularly broad suite of available technologies. Technology mandates are being used in the United States, with important variations. In California, regulation under development by the California Department of Toxic Substances Control (DTSC) as part of the California Green Policy Initiative requires that firms manufacturing “priority products” that contain “chemicals of concern” perform an alternatives assessment to identify whether there are any economically and technically feasible safer options. In cases where such alternatives exist, DTSC can decide to limit or completely phase out the use of the more hazardous chemical in order to promote adoption of the greener option. This, along with similar programs, including one in Maine, is an example of a more flexible technology mandate. Regulatory restrictions on the use of particular chemicals or on levels of emissions can be major drivers of green chemical innovation, without mandating the use of particular technologies. One example from China was driven by the Montreal Protocol on Substances that Deplete the Ozone Layer. In order to meet its commitments under the treaty, China closed down a number of small firms that produced ozone-depleting chemicals as a byproduct of their processes, while also working to develop alternative technologies for the remaining, larger firms. The government invested funds to support collaborative research and development among academia, the State Environmental Protection Agency (SEPA, now the Ministry of Environmental Protection) and industry, which ultimately led to a new, greener process.

Regulatory restrictions in green chemistry can be closely related to technology mandates, which, when accompanied by penalties, can create enough of an incentive—even to the point of “comply or go out of business”—to compel innovation. State-level strategies in Maine and California, and potential future regulation in the United States under bills proposed in the U.S. House of Representatives and Senate, include provisions to use regulation to limit or ban the use of particular chemicals. The advantage of a regulatory approach over prescriptive technology mandates in the case of green chemistry is that they allow for continuous adaptation as firms can determine the most profitable solutions for their specific context without being locked into particular technologies in cases where regulations fail to keep pace with changes in science and technology.

One important strategy to support green chemistry innovation is the provision of funding far basic and applied research. In the United States, there have been several attempts to pass a green chemistry bill at the federal level, which would direct the National Science Foundation and other agencies to allocate a portion of their funding budgets toward green chemistry. The most recent version, House 2850, was passed in the U.S. House of Representatives but did not come up for a vote in the Senate before the end of the legislative session in 2008. In China, professors and bureaucrats reported that large portions, ranging from 25 to 40 percent of the available research funding in chemistry and chemical engineering from the National Natural Science Foundation of China, are being earmarked for green chemistry work.

Green chemistry is an interdisciplinary endeavour, one that has benefited from a variety of collaborations and partnerships. One approach is the public-private partnership, which can have a multi-sectoral dimension. In these arrangements, stakeholders from government, industry, academia and civil society work together to accomplish technical and policy goals. The Green Chemistry Institute, which started in 1997, was founded with such collaboration in mind. Though based in the United States, GCI has built an international green chemistry network with chapters all over the world and hosts several different industrial Green Chemistry Roundtables. These roundtables have become important for companies to exchange pre-competitive knowledge, develop common tools and even commission key pieces of research.

Partnerships and collaborations are also being used in other parts of the world. In China, the government is actively promoting research programs that involve private firms working with academia. This builds on an existing structure where, historically, the Chinese domestic chemical industry looked to universities for help with the adoption of new technologies. Several leading green chemistry groups have long-standing relationships with their local chemical industries. The change in recent years has been an expansion beyond a model where universities do contract work with industry on specific chemical technologies, to the establishment of joint, long-term research programs focusing on a broader array of green chemistry technology. The goal of these collaborations is to improve communications between industry and academia in terms of research needs, as well as to help move promising innovations from the laboratory, through the pilot phase and into broader commercial adoption.

Government can also work directly with industry. In the context of green chemistry, this has taken the form of technical assistance for small and medium-sized businesses that could benefit from the implementation of green chemistry innovations. In many cases, these firms are either unaware of the existence of potentially beneficial technologies or lack the in-house expertise to implement them. In

Massachusetts, there is evidence that the Office of Technical Assistance has been particularly effective in helping firms implement technologies that allow them to meet the goals of the Toxics Use Reductions Act. Firms are attracted to many aspects of these innovations; not only are they better for the environment, but they can also significantly improve worker safety. For example, the use of super-critical carbon dioxide ([CO_2]) in dry cleaning can replace the highly toxic solvents currently in use. Other examples of areas that have benefited from green chemistry technical assistance programs include car painting and finishing, hair and nail salons, pulp and paper processing facilities and tanneries.

For the most part, there is very little, if any, statistical data that can be used to determine which of these ten approaches have been the most effective. However, empirical observation and interviews with individuals in the chemical industry reveal that certain patterns are emerging. Generally speaking, there are two major elements of successful innovation in green chemistry that policy can usefully address. The first is the process through which scientific and technical discoveries are developed and moved into practice. This process, covering not only development but also dissemination, includes support for collaborative, use-inspired, and basic research as well as support for projects aimed at lowering barriers to practical implementation of new technologies. The second element of successful innovation is policymaking that deals effectively with market failures and encourages the creation of new markets required for green chemistry innovations to be profitable.

Effective policy could encourage voluntary collaborations between firms and among government, NGOs and industry. Collaborative activity has been especially important in developing underlying scientific knowledge, tools and metrics, creating awareness and disseminating specific technologies. Green chemistry is a complex, interdisciplinary science and it appears to benefit from the involvement of experts from a variety of fields. It has also proven to be an area in which firms are willing to cooperate and share expertise. This has been the case with the different industrial roundtables of the GCI, as well as other independent industrial associations such as the Green Chemistry and Commerce Council.

Collaboration with government has been another factor contributing to the development of green chemistry. In both China and the United States, the government has had success as a facilitator of programs. The Chinese government has invested both time and resources in bringing industry and academia together to research, develop and implement particular technologies. In the United States, the government has collaborated with industry through programs like the Environmental Protection Agency's Design for the Environment, as well as other local technical assistance efforts. Collaboration is also seen at the state level: Michigan is facilitating the Green Chemistry Roundtable, which includes academics, industry representatives and NGOs from around the state, in order to jointly plan green chemistry activities in the state. These kinds of collaborative programs help improve the flow of knowledge and expertise throughout the system and support the dissemination of technologies across different groups of users. Policy can also have a dramatic impact on markets, the second element of successful innovation. Over the past few years, pressures from private actors within the supply chain have created market incentives for new green chemistry innovations, even in areas with weaker environmental regulation and enforcement. Much of the pressure has come not from consumers, but from retailers and consumer products companies that are more likely to be subject to government policy.

Interviews conducted within many influential firms indicate that progress occurs when there are champions both at the technical and managerial levels. This is mirrored at the industry level, where the involvement of a few early adopter firms provides a powerful demonstration of the viability of innovations in the market. Individual and corporate champions often view meeting environmental challenges as critical to the long-term survival of their firms, while also providing opportunities to reap market gains from new products or achieve efficiencies in the use of existing resources. Industry leaders in the use of green chemistry have developed metrics to measure their environmental performance and have set concrete corporate goals for improvement. S.C. Johnson & Son, Inc., a recipient of the U.S. Presidential Green Chemistry Award for its Greenlist program, is one example. In another case, Nike recently announced that it will start releasing the sustainability ratings of many of its footwear lines. The rating system has been operational internally for some time, but the results were not made publicly available. In these cases, internal tools are being turned into public marketing strategies.

For green chemistry, and for innovations for sustainable development more generally, there is a need for greater public intervention, including strictly enforced regulation, investment in basic research and education to build human capacity, enhanced outreach programs, collaboration with industry to aid with technology transfer and implementation, and economic incentives for firms that may have the desire but not the financial capacity to make use of these innovations. Although the role of supply chain actors is key in fostering innovations for sustainable development in green chemistry, as argued in this chapter, voluntary collaborations and the influence of major supply chain actors, on their own, are not powerful enough to catalyse the increases in scale that are needed for a real transition to sustainability. If green chemistry becomes the norm for how chemical products and processes are designed, the chemistry sector can, through the use of life cycle thinking, provide the foundation for many other key green technologies and products required to ensure that future growth and development is truly sustainable.

Recent years have seen a disheartening string of revelations in which everyday items once considered safe—food, packaging, toys, clothes, furniture, electronic components, and many more products—are found to contain carcinogens, endocrine disruptors, and other harmful chemicals.

Growing demand for healthier alternatives, already seen in food production and housing construction, is also happening at the building-block level of manufacturing, where so-called green chemistry represents a revolutionary change in preventing pollution and health problems starting at the chemical design stage. Many industry and government entities are beginning to espouse the principles of green chemistry on their websites and in public statements. Now comes the task of crafting policy to put those principles into action.

The U.S. Environmental Protection Agency (EPA) defines green chemistry as "the design of chemical products and processes that reduce or eliminate the use or generation of hazardous substances. Green chemistry applies across the life cycle of a chemical product, including its design, manufacture, and use." Green chemistry also aims to mitigate the type of uncertainty Alan Goldberg, a professor of toxicology at the Johns Hopkins Bloomberg School of Public Health, recently described to The New York Times: "I can get [toxicity] information on only 20 percent of chemicals we interact with on a daily basis." Of that 20%, he now says, he may be able to find information on overt toxicity for about half, but for details on specific effects such as developmental neurotoxicity, the figure shrinks toward zero.

So what does green chemistry look like? Consider the example of pregabalin, the active ingredient in the neuropathic pain drug Lyrica[R]. Pfizer developed an alternative green-chemistry process that converted several steps of pregabalin synthesis from use of organic solvents to water.

That reduced both health hazards and production hearing requirements. With the new synthesis, waste from the process dropped from 86 kg of waste per kg product to 17 kg, and energy use dropped by 82%. Proponents say that's how the field can offer a win—win—win solution: good performance, lower cost, and less environmental impact—what Richard Engler, program manager of the EPA Green Chemistry Program, calls the "triple bottom line." For many, a standard is a logical next step. "At some point you have to go beyond a definition and principles," says Engler. "I think that's something the standard will enable."

Crafting a Standard

The American Chemical Society (ACS), a congressionally chartered independent professional organisation, through its Green Chemistry Institute[R] (GCI), is working with federal agencies, nonprofits, end users, and industry to craft a "business-to-business" standard for measurably reducing hazardous materials in all kinds of products and processes. The effort, says ACS GCI director Robert Peoples, is very close to having a standard to be instituted by the American National Standards Institute (ANSI), a nonprofit organisation that has coordinated voluntary industry standards in the U.S. private sector since 1918. Peoples says the ACS aims to issue its draft for public comment by this summer.

Peoples compares the likely impact of an ANSI standard to that of LEED[R] (Leadership in Energy and Environmental Design), the professional standard for green construction for which the U.S. Green Building Council offers accreditation. An ANSI standard would motivate whole industries to reconfigure their production processes for lower environmental and health impacts, says Jim Solyst, a principal consultant with ENVIRON International Corporation.

Working in partnership with the standards development organisation NSF International, ACS GCI established a Joint Committee, which has been guided by two subcommittees and a balance of industry, nongovernmental, public health, academic, and government representatives. The standard-development process and meetings are open to anyone interested, says Peoples. He anticipates ACS GCI will roll out a finalised standard by late fall 2010, adding, "That's going from a blank sheet of paper to an ANSI standard in just under two years."

Like LEED, the ANSI standard will be voluntary. Companies will be able to confirm their adherence through one of several likely certification levels. The most stringent is independent third-party certification; below that is certification by a trade association or other party separate from the company. A voluntary standard, says Peoples, acknowledges that finding new ways of formulating many items won't happen overnight. "Voluntary standards allow the marketplace to drive the changes. It will take a long time, and there's no single path or approach," he says. By setting a high bar, however, a standard drives everyone to a higher level of performance, Peoples says.

The EPA has been a key federal agency partner in developing a standard since green chemistry was "born" there in the early 1990s (the current EPA assistant administrator for research and development, Paul Anastas, is one of the authors of the 12 principles of green chemistry (7). Others involved include the U.S. Department of Commerce, which has developed its Sustainable Manufacturing Initiative to support the implementation of sustainable manufacturing practices and explore its implications for U.S. global competitiveness and firm profitability, says Blandine Trouille, a senior trade analyst with the department.

Trouille notes that the September 2009 meeting of the G-20 (leaders from 20 countries that together represent 85% of the world's economy) "was all about sustainable economic development. That means manufacturing in a different way." With that emphasis in the global marketplace, she adds, "Having a standard helps an industry or company show they're not simply talking the talk, but walking the walk." And other stakeholders acknowledge that since information on environmental health problems is now more accessible, the marketplace itself demands solutions faster than ever before.

Greening the Defence Industry

Department of Defence (DOD) officials expect their own guidelines and policies, along with broader federal ones, will have more impact than an ANSI standard. The DOD wrestles with the particular needs of its national security mission, which can make chemical substitutions difficult, says Carole LeBlanc, special expert on chemical and material risk management in the DOD Office of the Deputy Under Secretary for Installations and Environment. Those needs include long-lived products, critical performance standards with little flexibility, and high-volume production, says LeBlanc, who leads the department's effort in green chemistry. "The mission of national security always wins in a tie [with other objectives]," says LeBlanc. But green chemistry principles can still come into play. For example, trichloroethylene (TCE), an organic solvent, is still used in many applications although it's a likely human carcinogen. When DOD staff proposed replacing TCE with methylene chloride, another solvent, LeBlanc weighed the options with an eye to uncertainty as well as known risk, recognising that regulations lag behind scientific knowledge. Methylene chloride may affect the nervous system but its role in human cancers is still inconclusive. In that instance, the DOD deferred replacing TCE. Explains LeBlanc: "We know exactly what TCE is going to do"—and therefore what precautions to take; the precautions for methylene chloride are less understood.

LeBlanc cites an earlier EPA program, the Significant New Alternatives Policy (SNAP) Program, as a model policy approach that identified best practices for selecting chemicals. SNAP prioritized problems and solutions: it first identified the easiest fixes and mandated their use, then identified harder issues for the medium and long term. "That set a road map for green chemistry," she says.

LeBlanc's office works through two channels: shaping policy through directives and instructions and through less formal interactions with the branches of the armed services. "The policy piece drives where the road and the rubber meet," she explains, but there are also "wonderful things you can achieve rather informally" through workshops across the service branches that reveal shared problems and possible solutions.

LeBlanc and colleagues identified best management practices for green chemistry that already existed at DOD. From there, they developed training curricula on acquisition systems for online courses at the Defence Acquisition University, which trains 126,000 DOD staff in acquisition, technology, and logistics. Those courses are in the final stages of review. "When you change a chemical, you almost always have to change a process," explains LeBlanc, and that requires training in the new process.

Also, after reviewing its existing programs for assessing chemical composition and safety, the DOD adopted a longer-term approach for tracking chemicals used by military personnel and contractors. A life-cycle assessment approach now examines products and chemicals through three phases of their use: acquisition and procurement, maintenance (the phase when most chemical exposures occur, says LeBlanc), and disposal. "This is a whole new way of looking at things for DOD," says LeBlanc. A June 2009 DOD workshop on reducing hazardous materials also supported the military branches in identifying problem chemicals and planning solutions. Since then, two similar workshops have taken place at the National Aeronautics and Space Administration and the Aerospace Industries of America, a contractors' group.

Carrots and Sticks

Among policies that promote green chemistry, the 2006 legislation by the European Union known as REACH (Registration, Evaluation, Authorisation and Restriction of Chemical Substances) is the most imposing. It addresses the human health and environmental impacts of all chemicals, mandates comprehensive information, and will be phased in over a decade. REACH will require companies to register chemicals they use with a new agency based in Finland. REACH applies to all chemicals produced in or imported into Europe.

It's too early to determine that law's health impact, but its impact on policy is already formidable. "REACH mandated the provision of information," explains Solyst, "a vast increase in the amount and availability of information on chemicals. That's good. And it has made downstream users of chemicals much mote aware." Before, he says, there was not sufficiently detailed information to make the public confident of the origin and composition of many chemicals used in daily life. REACH provides that level of information—but also creates a huge task in managing it.

For LeBlanc, REACH is in part an attempt to give Europe a competitive advantage in a global shift toward sustain-ability. "This is very much a trade issue for the United States," she says, adding it will have "a huge impact" on the DOD's supply chain and information in its Material Safety Data Sheets.

Stateside, federal government wide programs for greening procurement include the EPAs Environmentally Preferable Purchasing and the Green Procurement Program managed by the General Services Administration. One key incentive program the EPA manages is the Presidential Green Chemistry Challenge Awards, which began in

1996 during the Clinton administration. The awards are typically given annually to outstanding individuals and businesses responsible for innovations in green chemistry in five categories: academic research, small business, designing greener chemicals, greener synthetic pathways, and greener reaction conditions. The awards have raised awareness of green chemistry and fostered the idea that with innovation, you can reduce both environmental impact and cost, Engler says; he explains, "It's not a tradeoff between the two."

Others are less sure. Elizabeth Grossman, an investigative journalist and author of Chasing Molecules: Poisonous Products, Human Health, and the Promise of Green Chemistry, sees mixed motives in the chemical industry. "Every large chemical company is investing in green chemistry," she says. "At the same time, those companies have huge long-term investments in the chemicals being substituted—for example, phthalates—and their shift off the market will put a dent in the revenue from longstanding products."

Chapter 8

Geochemistry of Sedimentary

The Iberian Pyrite Belt is the world's greatest concentration of large massive sulphide deposits associated with volcanic rocks and hence the palacovolcanological setting has to be well documented if the ore-forming process is to be understood. Currently no consensus exists on the volcanological style of the Volcano-Sedimentary Complex host to this important accumulation of metals and two contrasting models arc being debated. Since the adoption of the exhalative model for the mineralisation, the host succession has been interpreted as a dominantly effusive, and explosively eruptive, volcanic pile. More recently this model has been challenged by the emergence of the sill-sediment-complex model in which up to 90% of the host rocks are high-level pcperitic intrusions. The two volcanological models lead to very different conclusions about the chronology of the host rocks, the timing and genesis of alteration, the driving force for the ore-related hydrothermal system, the source of metals, and synmineralisation reconstructions.

A distinctive feature of the Volcano-Sedimentary Complex is the alternation of primary volcanic facies and sedimentary facies. This presents an opportunity to discriminate between the competing volcanogenic models because some provenance patterns in the sedimentary intervals are expected to be very different in the two settings. Saez et al. (1996) expressed the commonly held view of the volcanology when they noted 'a predominance of sub-aerial or shallow-water volcanic activity' and claimed strong support for 'a prevailing explosive mechanism for the felsic volcanic activity'. Such a volcanic style would be expected to contribute much volcanic detritus to sedimentary intervals. In sill-sediment complexes there does not have to be any genetic link between the sills and overlying sedimentary rocks. In the extreme end-member case, represented by the still evolving complex in the Guaymas Basin, none of the igneous rocks have reached the surface and, because they are totally confined, cannot provide detritus to the sedimentary environment. Sediments between the mafic sills in the Guaymas Basin are a mixture of continent-derived gravity-flow deposits and oceanic pelagic deposits. The other end of the spectrum in sill-sediment complexes is represented by cryptodomes that have evolved to become largely emergent where much of the overlying sediment is a carapace breccia, derived from the locally exposed igneous body. Between these end-members lies what is probably the most common sill-sediment situation where the cover to some intrusive sheets breaks down and hydrovolcanic eruptions feed fluidized mixtures of fragments of the igneous rock, and the host sediments, into the depositional environment.

Intrusive sheets may, therefore, intrude entirely non-volcanogenic sediments and/or the products of earlier hydrovolcanism. In this latter case sills may or may not have chemistry comparable with that of the overlying sediments. This study documents the geochemistry, and interprets the provenance, of sedimentary rocks within the Pyrite Belt volcanogenie host sequence and compares these data with predictions made by the two competing volcanic models about the expected nature of sedimentary rocks interlayered with primary volcanic facies.

Tectonic Setting of the Iberian Pyrite Belt

The Iberian Pyrite Belt forms part of the South Portuguese Zone, a tcrrane separated from the Iberian Massif by the exotic, oceanic Pulo do Lobo ophiolitic terrane. The Volcano-Sedimentary Complex is a volcanogenic sequence that averages less than 1 km in thickness and is set in an otherwise terrigenous succession. Rhyolites and rhyodacites are volumetrically dominant in the Volcano-Sedimentary Complex but its overall character is bimodal with a second mode of basaltic rocks; intermediate compositions are widespread but minor. The underlying Phyllite-Quartzite Group clastic sequence (sandy turbidites and black mudrocks) formed in the Devonian at a passive margin.

For the metallogcny of the region, the key tectonic event was the transition, late in Devonian times, from this drift phase of passive margin evolution to active extensional tectonics. This switch was induced by oblique convergence as the South Portuguese Zone was sutured to the main Iberian Massif. Most workers have interpreted the Volcano-Sedimentary Complex magmatism as being a consequence of lithosphcric stretching during the oblique suturing event.

Towards the end of the Viscan, as the collision progressed, a switch from transtension to transpression resulted in deformation propagating southwestwards. Ahead of this deformation front there is a record of synorogenic sedimentation (flysch) that is referred to as the Culm. Regional metamorphism associated with the orogeny varies from very low to low grade, generally in a north-south gradient from zeolite facies in the south of the South Portuguese Zone to upper greenschist facies in the north all mudrocks in the region are now slates, but sedimentary facies can be safely identified and, throughout the discussion, terms relating to pre-metamorphic state will be used. Strongly sericitized volcanic rocks are also well cleaved but many macro-textures can still be recognised.

Contrasting Volcanological Models for the Volcano Sedimentary Complex: Rio Tinto District

Because the Rio Tinto district has been documented in detail in the context of both volcanological models, this region provides the best opportunity to determine which interpretation gives the most satisfactory explanation of the observed sedimentary provenance relationships.

According to the volcanic-pile model, the Volcano-Sedimentary Complex around Rio Tinto has a simple tripartite lithostratigraphy that is laterally continuous over tens of kilometres. This comprises a 400-500 m thick Basic Complex at its base, succeeded by a 50-800 m thick Acid Complex, and an uppermost Transition Scries, which can be several hundred metres thick but is commonly a few tens of metres. In this interpretation, the Basic Complex consists of basalts, pillowed in places, and derived volcaniclastic rock together with background mudrock. The Acid Complex has been described by Garcia as consisting of 'very homogeneous acid volcanics (dacitic to rhyolitic in composition), with predominant pyroclastics, and locally abundant breccias, autobrcccias, and lavas', and he described the Transition Series as being a 'heterogeneous pyroclastic level' comprising 'polygcnetic breccias, litho-crystalline tuffs, chloritized-haematizcd-scricitizcd tuffs, and massive sulphides'.

The following section summarises the findings of Boulter et al. (2001). In their sill-sediment-complex model, the primary volcanic facics are interpreted as predominantly subconcordant sills and more than 80% of the Volcano-Sedimentary Complex is a bimodal suite of post-mineralisation felsic and mafic pepcritic intrusions. It is proposed that this fold structure was the site of extensional growth faulting, which marked the end of Phyllite-Quartzite Group style sedimentation and controlled the formation of a wedge-shaped unit of debris-flow deposits. Earlier workers referred to this unit as the 'Slate Conglomerate', a distinctive interval of thickly bedded and poorly organised pebbly mudrocks and black mudrocks that ends in several metres of black mud. Apart from a few per cent of andesitic pebbles, the great bulk of the pebbles are muddy sandstones, muddy siltstoncs and anoxic mudrocks; this unit was overwhelmingly sourced in continental regions.

Towards the end of the phase of extensional tectonics, magmas utilised the extensional faults and were emplaccd in the vicinity of these structures mainly as composite basaltic andesite to rhyolite high-level intrusions of hundreds of metres thickness; these magmas were largely emplaced directly beneath the 'Slate Conglomerate' and its immediate lateral equivalents. The andesitic component became locally eruptive as hydrovolcanic eruptions that were generated where the cover to peperitic sills was breached. These eruptions deposited pebble-grade volcaniclastic breccias of an intermediate composition within a few kilometres of the Rio Tinto Anticline. At greater distances these breccias pass laterally into andesitic turbiditcs a few tens of metres thick.

The major hydrothermal event that created the Rio Tinto massive sulphide sheet was initiated during the extensional faulting phase, i.e. syn-'Slate Conglomerate', but peaked after the basaltic andesite-rhyolite magmatism that is pervaded by the stockwork feeder to the exhalative mineralisation. Waning hydrothermal flow resulted in chert deposition over some of the sulphide sheet. For the Rio Tinto district, the mineralisation was followed by a pronounced change in the redox state of the stratified sequence. Sedimentation characterised by radiolarian-bearing red mudrocks is typical of the interval between the sulphide layer and the flysch of the Culm. Intercalated with the red mudrocks are felsic volcaniclastic rocks in which very thickly bedded gravel-grade breccias are much more important than sandy facies. Most of these are debris-flow deposits directly or indirectly derived from hydrovolcanic eruptions associated with felsic peperitic intrusions, only one of which is preserved in the region. This intrusion, on the eastern edge, is a microphyric quartz-feldspar porphyry of 100 m thickness and 400 m strike length. The post-mineralisation red mudrock unit contains one of the greatest lateral facies contrasts in the region. The felsic volcaniclastic rocks overlying the sulphide sheets are thinly bedded (millimetres to tens of millimetres thick) sandy turbidites in a mudrock-dominated sequence. Where there are no sulphide sheets, felsic gravel-grade deposits predominate, and beds metres to tens of metres thick are common. This distribution defines an antithetic relationship between breccia facies in the red mudrock sequences and stratiform mineralisation.

Until late in the evolution of the marine red-bed sequence, the only primary volcanic facies in the region were the felsic microphyric intrusion in the red mudrock unit and the premineralisation intermediate-felsic igneous rocks in the Rio Tinto Anticline. Subsequently, a bimodal suite, which represents more than 80% of the volume of the Volcano-Sedimentary Complex in the area shown, intruded this assemblage. Outside the Rio Tinto Anticline (i.e. in non-mineralised regions) the bimodal suite intruded the 20-30 m of stratified andcsitic volcaniclastic rocks, interbedded with black mudrocks, and the overlying hectomctric red mudrock assemblage, as well as locally cutting down into the Phyllitc-Quartzitc Group. The typical section comprises one or two subconcordant felsic sheets of several hundred metres thickness and 5-10 mafic sheets each of which is typically several tens of metres thick. The bimodal package of felsic and mafic sills shows regionally discordant relationships. Westwards both the felsic and mafic sills reach progressively higher levels in the stratigraphy defined by the sedimentary rocks. This trend continues until in the Odiel River the mafic sills intrude the red mudrock-felsic volcaniclastic facies association. At this western locality, immediately below the main felsic sheet, in 420 m of stratigraphy, two main dolerite sheets averaging 135 m thick and a 20 m thick dolerite sheet alternate with red mudrocks and cobble-grade felsic volcanic-derived breccias.

Reference Measured Section

An almost fully exposed section through the Volcano-Sedimentary Complex in the Jarama River provides a reference section for discussion of the provenance relationships. The data presented here mainly come from, or can be reliably correlated with, the logged section and the mapping of Oswin (1986) allows this section to be tied. To emphasize the characteristics of the bedded rocks, the right- and left-hand columns exclude the primary volcanic facies. For the volcanic-pile model, the central column would reflect the chronology of the volcanic accumulation. For the sill-sediment-complex model, the marginal columns, if stacked, represent the sequence of stratified rocks before the intrusion of the bimodal suite.

Primary Volcanic Fades

Below the lowest sheet of largely coherent mafic igneous rocks is the Phyllite-Quartzite Group, a sequence of turbidites and black mudrocks. The lowest third of the Volcano-Sedimentary Complex comprises mafic sheets, averaging 30 m thick, intercalated with sedimentary rocks. This zone is equivalent to the Basic Complex of Garcia (1980, 1990). Interleaved with the mafic sheets are turbiditic sedimentary rocks in a background of black mudrock.

Petrographic and geochemical data on the mafic sheets support the field classification. Where plotted on a Zr/ TiO_2 v. Nb/Y diagram they fall in the basaltic andesite and subalkaline basalt fields. The mafic rocks have the most extreme Ti/Nb ratios of all the igneous rocks in the district, being in excess of 980:1. Petrographically the mafic sheets are dominantly pyroxene and plagioclase in near equal proportions. The feldspars are randomly oriented and subophitic textures arc common. Grain sizes are commonly 0.5-3 mm.

The three thick felsic units in the logged section are distinctive because of their lack of stratification; there are only 10 m of sedimentary rock in these 300 m. This felsic interval is equivalent to the Acid Complex of Garcia (1980, 1990). The felsic sheets in the Jarama River section are quartz-feldspar porphyries and they all plot in the rhyodacite-dacitc field of the Zr/TiO_2 v. Nb/Y diagram. In the Jarama section, all of the components of the felsic sheets have Ti/Nb ratios less than 400:1, and most are less than 300:1.

Sedimentary Fades

In the black mudrock interval, most of the sandy turbidites are medium sand grade and are mainly composed of T_a and T_b divisions of low-density turbidites; other divisions are rare. For the sandy beds logged in this portion of the Jarama section, the mean bed thickness is 4.96 cm, ranging from 2 mm for a fine-sand layer to 44 cm for a turbidite that has granules at its base.

Modal analysis of the framework grains has been carried out on the least deformed samples. The most striking feature of the volcanic turbidites associated with black mudrock is the virtual absence of quartz. The framework population of sample 8 is lithic clasts and heavily altered feldspar. More than 80% of the framework of sample 9 is lithic clasts, all similar in character. The rock fragments are mainly highly feldspar-phyric (quartz absent) with feldspar lath concentrations up to 50% set in a trachytic-textured groundmass. Rare aphyric clasts have, as their only recognisable grains, crowds of strongly aligned feldspar microlites c. 0.02 mm long. Above sample 9, in the uppermost final 5-10 m of black mudrock stratigraphy, lithic wackes contain phenocryst-rich clasts that have a minor (<2%) quartz-phyric component. Vesiculation of the clasts was very limited and clasts are typically bound by planar to slightly curviplanar fractures.

Above the felsic sheets in the Jarama section, there is a section of 70 m thickness dominated by felsic volcanic-derived gravelgrade volcaniclastic deposits where mudrocks are red (left-hand column, Fig. 4). This unit correlates with the lower part of the Transition Series of Garcia (1980, 1990). This gravel-dominated interval is largely thickly bedded, poorly organised sedimentary breccias. These beds commonly range from 5 to 15 m thick although the maximum is 32 m. These depositional units are composites of crudely graded sedimentary breccias at their bases and laminated pebbly sand facies in their upper portions. The boundary between the graded and laminated parts of the beds is locally marked by the presence of outsized clasts several tens of centimetres across compared with the centimetre-sized clasts below. Such a concentration of outsized clasts is believed to occur at the boundary between a lower traction carpet, which creates a debris-flow deposit, and an overlying turbulent cloud, which creates the laminated facies. It is possible that flow transformation within the body of the debris flows could have led to the development of turbulence, resulting in the uppermost laminated facies of these beds. The laminated facies commonly occupies around 20% of a bed but, in some cases, is found independently of the graded breccias.

There is considerable textural variation in the graded breccias, many of which consist of angular to subangular cobbles with rare outsize clasts up to several metres in diametre. Most beds are almost monomictic, comprising quartz-feldspar porphyry clasts, whereas some other beds are highly heterogeneous. The most admixed clast populations include felsic rocks with various phenocryst populations (quartz and/or feldspar in varying size and abundance), and sedimentary clasts such as red mudrock, chert and manganiferous chert. Clasts arc very rarely amygdaloidal and the generally dense (non-vesicular) character of the igneous clasts is a distinctive feature. Several per cent of the clasts are fragments of peperite.

Sedimentary Geochemistry

Major elements and selected trace elements were determined by XRF, following standard procedures, on a Phillips PW1400 instrument at the Southampton Oceanography Centre, using a 3 kW Rh-anode tube. Analytical precision is better than 5% (2 SD) for the major elements and 5% (2 SD) for the trace elements. Precision and accuracy are monitored using international standards analysed as unknowns. The REE were analysed by inductively coupled plasma mass spectrometry (ICP-MS) at the Southampton Oceanography Centre on a VG Plasmaquad II instrument. Precision is better than 4% (2 SD). In each sample, run precision was determined by using the international rock standard BHVO as the drift monitor after every five unknowns. A series of international rock standards was used to determine accuracy.

Figure: *Elemental Geochemistry of Sedimentary Rocks*

Petrogenetic studies on the igneous rocks of the Pyrite Belt have documented and discussed the nature of the non-stratified rocks. Here we consider the sedimentary rocks in the Volcano-Sedimentary Complex for which previous geochemical studies have tended to concentrate on the non-clastic components in the sequences that host the manganiferous deposits. The sedimentary geochemical data presented here have been organised into two main stratigraphie groupings according to the colour of their associated mudrock facies: black (lower) and red (upper). Within these, further grouping is based on sedimentary facies; namely, mudrock facies, sandy turbidites and gravel-grade sedimentary breccias.

The major- and trace-element chemistry for the stratified rocks is complex, as is shown by the lack of a single linear array on two-component Harker-type diagrams. Overall major-oxide concentrations show large ranges in weight per cent (SiO_2 92.75-59.55; Al_2O_3 20.28-2.75; Fe_2O_3 25.71-2.03; MgO 5.89-0.5; CaO 10.24-0.03; Na_2O 6.37-0.15; K_2O 5.71-0.24). However, there is a greater degree of consistency in the major oxide contents within the facies categories (black mudrock, red mudrock, sandy turbidites and sedimentary breccias associated with black mudrocks, and sandy turbidites and sedimentary breccias associated with red mudrocks). For example, mean MgO and CaO concentrations in the sandy turbidites and sedimentary breccias associated with the black mudrocks are around twice the level of those in sand- or gravelgrade sedimentary rocks associated with the red mudrocks. The values for Na_2O and K_2O are comparable for the coarse volcaniclastic rocks in both the red and black mudrock portions of the stratigraphy.

Overall concentrations of the ferromagnesian element Ni are very low, averaging 25 ppm. Although the concentrations of most high field strength elements are not particularly high (e.g. Nb average is 11 ppm, Th average 7 ppm), Zr in contrast, in the coarser of the stratified rocks intercalated with the red mudrocks, is abundant, averaging 282 ppm. Y concentrations discriminate between the oxic and anoxic sedimentary facies, being higher in red mudrocks and associated facies (average 46 ppm) than in black mudrocks and related coarser facies (average 20.3 ppm).

The clearest discrimination of the facies groups is shown by plots of $ZrATiO_2$ v. Nb/Y. Sandy turbidites, intercalated with the mafic sheets, virtually all fall in the andesite field. All the sand- and gravel-grade rocks, which are interbedded with red mudrocks, plot well within the rhyodacite-dacite field as is to be expected from the approximately equal quartz and feldspar phenocryst distribution in the clasts. Nb/Ti plots for the sand-and gravel-facies rocks show that the distribution of these elements even more clearly segregates samples in accordance with their petrography. The sandy turbidites and gravel-grade sedimentary rocks interbedded with the black mudrocks have Ti/ Nb ratios between 400 and 800 whereas the sand- and gravel-grade sedimentary rocks associated with the red mudrocks all have Ti/Nb ratios below 400. Additional data for the red mudrocks, and interbedded coarser clastic rocks, come from the trace-element lithogeochemical survey by Oswin (1986) of the Transition Series. His data for these sedimentary rocks show a very clear restriction in Ti/Nb ratios to less than 400:1.

Red and black mudrocks from the Volcano-Sedimentary Complex, along with a sample from the Phyllite-Quartzite Group, occupy the same general field on Ti/Nb plots, consistently having Ti/Nb ratios less than 400:1. Data from the Phyllite-Quartzite Group presented by Mullane (1998) show that material from this stratigraphic unit always falls in this field, being mainly restricted to Ti/Nb ratios between 200:1 and 400:1, and that Nb is generally abundant (most > 13 ppm). Similarly, the chondrite-normalised REE patterns of all the mudrocks essentially have the same shape with light REE (LREE) enrichment, flat heavy REE (HREE) segments, and significant negative Eu anomalies. (La/Lu)N ratios range from 6.49 to 10.31, except for sample 32, which has a ratio of 2.83. All of the REE analyses have Eu/Eu* ratios in a narrow range of 0.56-0.74. Some differences are also noteworthy. Four samples have almost identical patterns. This group comprises sample 1 from the Phyllite-Quartzite Group several hundred metres below the Volcano-Sedimentary Complex; sample 22, a slate within the pebbly mudstones of Corta Atalaya ('Slate Conglomerate'); sample 28, from a stratified zone low in the thick felsic sheet SE of Rio Tinto; and sample 35, a hematitic slate c. 10 m above the massive sulphide layer in Corta Atalaya. Samples 20 and 23, intercalated with the andesitic turbidites in the Jarama River section, stand apart from the grouping of four noted above because of their lower total REE abundances. However, all six of these samples have comparable patterns, although samples 20 and 23 show lower LREE to HREE fractionation as illustrated by $(La/Lu)_N$ ratios of 6.65 and 6.84. The two remaining analyses are distinctive. Sample 29 has the lowest REE total and sample 32 has lower fractionation between LREE and HREE, which is reflected in a $(La/Lu)_N$ ratio of 2.83.

Plots of $ZrATiO_2$ v. Nb/Y and Ti v. Nb data for the primary volcanic rocks facilitate comparison with the associated sedimentary rocks. The igneous rocks have been subdivided into pre- and post-mineralisation groupings based on the field criteria established by Boulter (1996). The pre-mineralisation igneous rocks all come from the Rio Tinto stockwork, although they can be classified using elements relatively insensitive to alteration. It is evident that the spectrum of igneous compositions emplaced before the major exhalative episode at Rio Tinto ranged from rhyodacite to basaltic andesite. The post-mineralisation igneous rocks are distinctly bimodal with a conspicuous absence of intermediate compositions.

Ascertaining provenance of sedimentary rocks is commonly frustrated by a multiplicity of processes that operate on the materials between source and sedimentary rock, including chemical weathering, hydraulic sorting, diagenesis, metamorphism and hydrothermal alteration. Fortunately, for the coarse clastic rocks considered here, the predominance of rock fragments, having a limited range of compositions and textures, means that there will be a much closer relationship between sedimentary composition and the source than for materials that have undergone more extended sedimentary sorting, or chemical weathering before any transport. However, identifying the chemical signatures of the parent to the mudrocks is a much more challenging process because of the multiplicity of chemical-mineralogical modifications associated with the evolution from fresh source rock to lithified mud.

Chemical modification is intense in this process and only a limited range of elements, generally those with low to moderate seawater-upper-crust partition coefficients, are capable of reliably reflecting the source composition. For some elements within this selection the possibility of quantitative transfer from source to sedimentary rock is reduced by factors such as heavy mineral fractionation, changes of solubility with differing oxidation states, and chalcophile behaviour. For these reasons geochemical provenance studies typically concentrate on the Tigroup, the Al-group, the REE, Th, Nb, Sn and be as the most reliable means of deducing the nature of source terrains.

Bonjour & Dabard (1991) have shown that Ti/Nb ratios are particularly good provenance indicators in clastic sedimentary rocks because these elements undergo little fractionation during weathering or diagenesis. The utility of Ti and Nb in sedimentary provenance studies has also been demonstrated by Slack & Stevens (1994). Recent work has confirmed the lack of mobility of Ti and Nb during weathering but has shown, in some circumstances, significant mobilisation of Y and some perturbation Of Zr.

Major Elements

Based on the considerations discussed above, only generalisations can be made when using the bulk of the major-element chemistry. The lack of single linear arrays on two-component plots indicates a complex picture for the major elements. A linear pattern on element-element plots for sedimentary rocks would typically represent mixing lines generated by the sorting of clay-rich and quartz-rich components during sedimentation, given a reasonably homogeneous source. If this process does not involve gain or loss of material then the array will pass through the source composition. For the Rio Tinto area, the lack of a well-defined correlation between pairs of major elements could imply significant clement mobility during any one or some combination of the overprinting processes noted earlier. Alternatively, or additionally, several sources may be implicated. Despite the potential for chemical disturbance during alteration, a mixture of felsic to intermediate provenance characteristics for the major-element chemistry is evident, based particularly on MgO, and K_2O, concentrations.

Contrasts between Igneous and Overlying Sedimentary Rocks

Petrographic and geochemical contrasts between sedimentary rocks and intercalated, or stratigraphically lower, primary volcanic facies are common in the Volcano-Sedimentary Complex. Of these patterns, the alternation of intermediate and mafic signatures, through at least five sedimentary-rock-dolerite-sheet repetitions in the Jarama section, is one of the most prominent. Ti/Nb ratios, and to a lesser extent Zr/TiO_2 v. Nb/Y data, clearly discriminate between the mafic sheets and the intercalated stratified rocks. Sedimentary packages with average Ti/Nb ratios of 512:1 alternate with mafic sheets that have average Ti/Nb ratios of 1142:1. It is difficult to explain these pronounced contrasts in terms of volcanic-pile processes. If the mafic sheets were effusive and associated with primary volcaniclastic rocks, they should have contributed detritus to any immediately overlying sediments.

Even in volcanic settings with subdued topography, such as mafic submarine lava plains, examples have been documented where intercalated sedimentary horizons have been entirely sourced in the mafic rocks. In comparable settings, the provenance of intercalated sedimentary rocks is not exclusively the local volcanic rock but its geochemical signature is clearly identifiable. The likelihood of local contributions to sedimentary provenance is increased by the recognition that deep-water volcaniclastic deposits are much more extensive than previously recognised and that deep-marine current activity is a significant contributor to the nature of these deposits. For the Jarama River section, the mafic sheets did not supply detritus to the overlying sedimentary rocks, as is demonstrated by the petrographic and chemical contrasts. The volcanic-pile model is not supported by these data, which instead favour intrusive behaviour for the mafic sheets as implied by the abundant field observations of widespread interaction between magma and wet sediment. Stratified rocks of a mafic composition are not recorded in the region and it would appear that the peperitic mafic intrusions did not break their cover to contribute to the sediment source.

Figure: *Igneous rock poking through a layer of overlying sedimentary rock beds.*

Additional information on the part of the Jarama section that contains the mafic sheets comes from the REE data on the mudrocks (samples 20 and 23) associated with the andesitic turbidites. These data, in pattern and absolute abundances, closely match REE data from Iberian Pyrite Belt andesites.

Basalts and basaltic andesites in the Iberian Pyrite Belt have flat REE patterns and do not have a negative Eu anomaly. This is very different from the patterns in the mudrocks intercalated with the mafic sheets in the Jarama section. Therefore, the mudrocks between the mafic igneous sheets have andesitic characteristics and these samples further highlight the lack of mafic rock provenance, thus supporting the field textural evidence for an intrusive mode of emplacement for the mafic sheets.

A more pronounced contrast, between the primary-volcanic-facies substrates and overlying stratified rocks, is represented by the alternation of mafic sheets and felsic volcaniclastic rocks in the Odiel River. Here the primary mafic volcanic facies alternate with felsic conglomeratic facies. The cobble-grade material in these sedimentary rocks is quartz- and feldspar-phyric and the bed characteristics arc comparable with the debrite zone towards the top of the Jarama section. The mafic sheets have not contributed at all to the felsic volcanic-derived volcaniclastic breccias, thus supporting an intrusive origin for the mafic volcanic rocks.

Geochemical data for the felsic igneous sheets within the post-mineralisation bimodal suite and the coarse clastic rocks in the red mudrock part of the stratified sequence show considerable overlap in their range of compositions. It is therefore not possible to use the geochemistry obtained in this study to link particular felsic sheets to their derived hydrovolcanic eruption products. Textural characteristics are more distinctive than chemical variations, but mingled magmas, and the composite nature of some sheets, limit the reliability of correlations in sections where both primary and stratified facies are felsic.

Persistence of Continental Signatures Through the olcano-Sedimentary Complex

REE data from mudrocks contribute to the identification of continental-derived components within the Volcano-Sedimentary Complex. The cluster of four nearly identical mudrock REE patterns is indistinguishable from the post-Archaean shale composites presented by Taylor & McLennan (1985) and considered by them to represent the upper continental crust for the REE although with a slight overestimate of absolute abundances. The (La/Lu)N ratios for this group range from 9.5 to 10.31 and compare well with the value of 9.2 for this ratio for the Post-Archaean Australian Shale of Taylor & McLennan (1985). All of these mudrocks would appear to be representative samples of average upper continental crust. Such a conclusion was to be expected for the Phyllite-Quartzite Group sample 1 because this is stratigraphically well below the Volcano-Sedimentary Complex.

In contrast to sample 1, samples 22, 28 and 35 all are located well within the Volcano-Sedimentary Complex, and their upper-crustal affinity poses considerable difficulties for the volcanic-pile model because volcanic substrates should contribute significantly to the overlying clastic rocks. The mudrock sample from the 'Slate Conglomerate' in Corta Atalaya (sample 22) is one of the closest matches to the Post-Archaean Australian Shale, yet, in the volcanic-pile model, it had formed after one-third to one-half of the Volcano-Sedimentary Complex. The REE data coupled with the pebble compositions create the most pronounced contrast between primary volcanic facies and overlying stratified rocks in the region. Schermerhorn invoked exotic explanations by describing them as 'a flow tilloid deposited from a mudflow descending into and coming to rest in a deeper part of the basin, among volcanics'. In the sill-sediment-complex scenario, the sequence of pebbly mudstones is viewed as forming before and partially overlapping the first peperitic intrusions in the region such that volcanic contributions were minimal, and therefore intercalated mudrocks would not have detectable volcanic chemical signatures. Volcanic textures in the andesitic to felsic rocks have largely been obscured by the stockwork-related alteration but preservation of some peperitic textures supports an intrusive origin.

Although sample 28 would be considered in the volcanic-pile stratigraphy to be above the Basic Complex and to be in the lower part of the Acid Complex, its composition closely matches the post-Archaean shale composites. It is inferred that the increase in thickness has meant that the sill has cut to deeper levels than elsewhere into the pro-intrusion Phyllite-Quartzite Group and included rafts of that unit. This accounts for the Post-Archaean Australian Shale-type composition of sample 28. Other characteristics of the adjacent stratified rocks, such as intraformational conglomerates and pre-lithification deformation, match the features of the uppermost Phyllite-Quartzite Group.

The upper-crustal REE pattern from the red bed sequence in Corta Atalaya (sample 35) is highly anomalous in the volcanic-pile model because it is part of the Transition Series that post-dates the bulk of the Volcano-Sedimentary Complex (the Basic and Acid Complexes) and would therefore be expected to show a volcanic signature. Sample 35 is intercalated with sandy turbidites on the several millimetres to few centimetres scale that comprise less than 10% of this part of the succession. As is the case for the other oxic-facies volcaniclastic rocks, the arenites are felsic. Volcaniclastic input into the sequence above the sulphide sheet appears to have been much less than in lateral facies adjacent to the mineralised area and this is mirrored in the mudrock chemistry. Because the primary volcanic facies that predate the mineralisation are restricted to the immediate vicinity of the sulphide deposits, it is possible that a positive topography had been created in these areas by the time exhalative activity had ceased. Deformation of the sea floor above the high-level intrusions, spatially related to the extensional faults, may have reversed the tectonic depression of the basin floor during the formation of the half-graben. The large debris flows associated with the red mudrocks appear to have been restricted to the lower-lying areas that had no, or minimal, primary volcanic facies until after mineralisation. Comparable hematitic mudrocks that cap the massive sulphide bodies at Aljustrel also have no volcanic aspects to their geochemistry.

Sample 32 is a red mudrock from the Jarama section that is laterally equivalent to the sample 35 locality. This sample is from the part of the section where debris-flow deposits dominate and the volcanogenic input can also be seen in the mudrock REE pattern, which is comparable with data published by Mitjavila et al. (1997) for dacites in the Iberian Pyrite Belt. Where the red mudrock succession is dominated by major volcaniclastic input as evidenced by beds up to 30 m thick, the mud may have been largely generated during the hydrovolcanic process, possibly by spalling of glassy material as fluidized sediment came into contact with cooling igneous material along fracture planes. In contrast, the mineralised zones, above the topography created by the pre-mineralisation igneous rocks, were not affected by this catastrophic sedimentation. These mud-dominated areas received only a little volcanogenic detritus in the form of very thin turbidites, and the background pelagic sedimentation was of a regional, continental, provenance as shown by the Post-Archaean Australian Shale-type REE data.

The Volcano-Sedimentary Complex contains many examples of pronounced compositional contrasts between stratified rocks and underlying primary volcanic facies. Particularly striking are the contrasting compositions between alternating, largely coherent, mafic sheets and volcaniclastic turbidites-debrites. In the Jarama River the mafic sheets alternate with stratified rocks of andesitic composition where distinct populations of Ti/Nb ratios highlight the chemical contrasts. In the Odicl River, the mafic sheets alternate with bedded volcaniclastic rocks of a felsic composition. Perhaps a more significant contrast is found in Corta Atalaya between the virtually non-volcanic pebbly mudstones stratigraphically overlying hundreds of metres of intermediate, and felsic, primary volcanic rocks. These relationships cannot be reconciled with the provenance patterns expected for a continuously evolving volcanic pile where the underlying primary volcanic facies should be available to contribute to any sedimentary horizon formed in a volcanically quiescent period. REE geochemical signatures of a continental style are distributed throughout the Volcano-Sedimentary Complex stratigraphic column, supporting the proposal that much of the magmatic activity was confined, i.e. not eruptive. That this was not exclusively the case is demonstrated by the volcaniclastic turbidites and debrites of intermediate and felsic compositions. The andesitic component of the stratified rocks was derived from localised areas of magmatism centred on zones that became sites of exhalative hydrothermal activity. In contrast, the mafic part of the post-mineralisation bimodal intrusive suite does not appear to have contributed to any of the stratified volcaniclastic rocks in the Rio Tinto district, and appears to have been totally confined. The test provided by the sedimentary provenance supports the field evidence for a major role for peperitic intrusions in the facies architecture of the host to one of the world's most significant mctallogenic provinces.

Confirmation of the sill-sediment-complex nature of the Volcano-Sedimentary Complex highlights the need to reassess the physical volcanology of this volcanic unit to provide a basis for the ore-deposit modelling. Reconstructions of the synmineralisation setting will be in error if they arc based on an inappropriate model for the physical volcanology. Other aspects of the metallogeny affected include the modelling of the hydrothermal system, the understanding of the alteration patterns, the determination of metal sources, mechanisms of ore-body segmentation, correlating the ore-equivalent horizon, and discriminating pre- and post-mineralisation igneous rocks. All of these topics also have relevance for exploration methods. It is vital that the physical volcanology of this region is well defined, but it seems certain from data presented here that the volcanic-pile model is invalidated and there is strong support for the sillsediment-complex interpretation.

Geochemistry of Coeval Mesozoic Plutonic

High precision U-Pb zircon age-dating of volcanic and plutonic rocks in Hong Kong (herein the area of the Hong Kong Special Administrative Region) has placed tight constraints on the timing of Jurassic to Cretaceous magmatic events in a small area (1050 km^2) of the SE China igneous province. The data reveal four discrete episodes of magmatism, generally of less than one million years duration, separated by longer periods of volcanic and plutonic quiescence. Such precise dating, together with detailed field, petrographic, geochemical and isotopic data, provides an opportunity to examine the close interrelationships between volcanism, plutonism, and tectonism. This chapter describes the compositional variation through time of Hong Kong volcanic and plutonic suites and discusses aspects of their petrogenesis and tectonic setting.

Geological Setting

Destructive plate magmatism along the Eurasian continental margin during the Mesozoic Yanshanian Orogeny culminated in voluminous intermediate to silicic eruptions and the emplacement of numerous, high-level granitoid plutons. These Jurassic to Cretaceous volcanic and plutonic rocks crop out over a broad (400 km) belt along the coastal region of SE China and record a complex tectonomagmatic history.

The coastal region of SE China is dominated by several large-scale NE-trending fault zones the most prominent of which is the Lianhuashan Fault Zone extending for over 200 km through the maritime provinces of eastern Guangdong, Fujian and Zhejiang. In eastern Guangdong, the Lianhuashan Fault Zone comprises a 30 km wide zone bounded to the north by the Shenzhen Fault and to the south by the Haifeng Fault. Numerous volcanic centres and plutonic assemblages within this zone, including those of Hong Kong, suggest that the Lianhuashan Fault Zone was a major locus of magmatic activity throughout the Late Mesozoic. This is corroborated by Sr and Nd isotope studies of the granitoid rocks, geophysical studies and zircon inheritance studies, which collectively endorse the existence of a deep crustal boundary through Hong Kong separating dominantly late Archaean crust in the NW from Proterozoic crust in the SE.

Approximately 85% of the area of Hong Kong is formed from Mesozoic volcanic-plutonic assemblages. The bulk of these rocks are of Mid-Jurassic to Early Cretaceous age and are divided into two volcanic groups and two granitoid suites. During the course of this study it has been necessary to introduce a refinement of this suite and group nomenclature in order to discuss the close compositional and temporal relationships among the lithological units. Accordingly, the granitoids are grouped in terms of magmatic pulses where a 'Pulse' is defined as a magmatic event in which one or more single intrusive units (plutons) with similar bulk chemistry and mineralogy were emplaced at the same time. Similarly, the volcanic units are grouped in terms of phases where a 'Phase' is the volcanic equivalent of a 'Pulse'. A period (or episode) of magmatic activity may comprise one or more pulses or phases.

Volcanic and Plutonic Associations

The Middle Jurassic-Lower Cretaceous volcanic and plutonic rocks of Hong Kong are both spatially and temporally related. Subvolcanic dykes, stocks, and plutons also commonly have whole-rock geochemical signatures similar to those of the volcanic formations making it possible to link compositionally the plutonic pulses to one or other of the volcanic phases. The broad compositional similarities between volcanic rocks and granitoids from the four main episodes of magmatism are demonstrated on a plot of TiO_2 versus Zr. However, within each of these episodes, the relationship between volcanic phases and plutonic pulses is by no means straight forward and not all intrusions have subvolcanic equivalents and vice-versa.

Volcanic rocks of the Tsuen Wan Volcanic Group and granitoids of the Lamma Suite are mostly exposed in the north and west of Hong Kong. Based on U-Pb zircon age-dating and whole-rock geochemistry, the Lamma Suite granitoids can be divided into two groups, here termed Pulse 1A and Pulse 1B. Pulse lA granitoids were emplaced between 164.6 and 161.5 Ma and are temporally and geochemically related to the volcanic rocks of the Tsuen Wan Volcanic Group (Phase 1). Granitoids of Pulse lB have yielded ages between 160.4 and 159.3 Ma but do not have a recognisable volcanic equivalent. In terms of whole-rock geochemistry, Pulse IA granitoids have distinctive I-type characteristics whereas those of Pulse IB have A-type characteristics. Based on the new U Pb age data, two granitoids previously assigned to subgroup 1 of the Lion Rock Suite are now included with Pulse IB of the Lamma Suite.

A significant break (13 Ma) exists between the first and second episodes of Mid-Jurassic-Early Cretaceous magmatism. Volcanic and granitoid rocks belonging to this second episode were emplaced between 146.6 and 146.2 Ma. They are designated as Phase 2 and Pulse 2 respectively, and are largely restricted to a 6 km wide, NE-trending zone through central parts of Hong Kong. These rocks comprise an assemblage of dominantly I-type extrusive and intrusive units, but include a group of late stage A-type plutons and dykes for which there are no known volcanic equivalents. Two compositionally distinct, intercalated, volcanic plutonic assemblages have been identified within the third episode of magmatic activity which took place between 143.7 and 142.7 Ma. These comprise a large volume of dacite to rhyolite volcanic rocks and I-type monzogranites and quartz monzonites (Phase 3/Pulse 3), and a smaller volume of trachyte to rhyolite volcanic rocks and associated quartz monzonite intrusions (Phase 4/Pulse 4). Both assemblages are exposed mainly in the south and east of Hong Kong and the granitoids have distinctive I-type compositions.

Volcanic and granitoid rocks belonging to the fourth main episode of magmatism (Phase 5/Pulse 5) were emplaced between 140.9 and 140.4 Ma. They are largely concentrated in the southeast of Hong Kong and comprise thick, caldera-related silicic tuff sequences, two large monzogranite plutons and several smaller quartz monzonite stocks and dykes. The volcanic rocks are dominantly rhyolite in composition and the granitoids vary from fractionated I-type compositions to A-type compositions.

The final pulse of Mid-Jurassic-Early Cretaceous silicic magmatism is recorded by a number of small quartz monzonite to monzogranite stocks and dykes in the south of Hong Kong. These rocks intrude monzogranite of Pulse 5 but have not yet been radiometrically dated. They are volumetrically minor by comparison with the main plutonic assemblages and may in fact belong to a late intrusive stage of the fourth episode. These rocks are tentatively assigned to Pulse 6 and have no known volcanic equivalents.

General features of the mineralogy and petrography of individual formations and plutons are described elsewhere and only a brief summary is necessary here. The Tsuen Wan Volcanic Group (Phase 1) is composed mainly of coarse ash crystal tuff although significant volumes of tuff-breccia, lapilli tuff. volcaniclastic sedimentary rocks and fine ash crystal tuff are also present.

Coarse ash crystal tuff is the dominant lithology of the Repulse Bay Volcanic Group (Phases 2-5) but significant volumes of fine ash vitric tuff, which is commonly welded, is also present and form thick ignimbritic sequences. Intercalated volcaniclastic sedimentary rocks have been recognised in many units of the Repulse Bay Volcanic Group whereas lava flows are relatively uncommon. The mineralogy of the volcanic rocks closely resembles that of the related plutonic rocks.

The I-type Lamma Suite (Pulse IA) granitoids are commonly K-feldspar megacrystic and medium grained, and contain strongly zoned plagioclase, with quartz, glomerocrystic biotite, and minor amphibole. Accessory minerals include magnetite, zircon, zoned allanite, and apatite. A-type granitoids (Pulse 1B) comprise fine grained biotite monzogranite and leucogranite the latter containing fluorite, sparse muscovite, and rare garnet. Rhyodacite and rhyolite porphyry dykes of Pulse 2 (Lion Rock Suite) contain large (5 20 mm) euhedral K-feldspar megacrysts along with subordinate bipyramidal quartz. zoned plagioclase, and biotite glomerocrysts set in a fine-grained granophyric matrix. Accessory minerals are similar to those of the Pulse I I-type granitoids. Equigranular biotite monzogranite is the dominant lithology of Pulse 3 and Pulse 5 granitoids. Plagioclase in these rocks is strongly zoned, and accessory minerals include magnetite, zircon, titanite, allanite and fluorite. Quartz monzonites of Pulse 3 and Pulse 4 are dominantly porphyritic and medium grained. The main mafic minerals are hornblende and biotite, and accessory minerals include zircon, apatite, and magnetite. Quartz monzonites and monzogranites of Pulse 6 are principally K-feldspar porphyritic and fine grained. Biotite and amphibole are the dominant mafic constituents while the accessory minerals include magnetite, zircon, titanite and fluorite.

Major and trace element composition on a plot of K_2O v. SiO_2, almost all of the analysed rocks fall within, or above, the high-K calc-alkaline field of Le Maitre (1989). Many samples are extremely K-rich and plot in the field for fractionated alkaline shoshonitic series compositions.

Phase I rocks show slightly increasing K20 abundances with increasing degree of differentiation whereas the rocks of Phases 2 to 5 show flat, or slightly decreasing abundances with increasing SiO_2 content. Some of the anomalously high K_2O contents shown by Phase 2 volcanic rocks correspond to those which fall outside the normal igneous spectrum. These K_2O enriched rocks may have been affected by low temperature hydration and devitrification of rhyolite glass.

Granitoids of Pulses 1A and 1B show similar K_2O abundances and fractionation trends to volcanic rocks of the Phase I with compositions plotting within the zone bounding shoshonitic and high-K calc-alkaline series compositions. K_2O contents and fractionation trends for Pulse 2 rocks are also similar to those of the least-altered Phase 2 rocks, and the same can be shown for granitoid and volcanic rocks of Pulse 3/Phase 3. There are small differences in the K_2O contents and fractionation trends between Pulse 4 granitoids and Phase 4 volcanic rocks, with the volcanic rocks showing a trend of slightly increasing KO with degree of differentiation compared to the slight decreasing trend shown by the granitoids. The variation in K_2O abundances in Pulse 5 granitoids is similar to that of Phase 5 volcanic rocks. Pulse 6 granitoids show slightly higher K_2O contents than Pulse 5. Overall, K_2O contents for rocks of Phases 2 to 5 and Pulses 2 to 6 are higher than for the Phase 1 and Pulse 1 rocks and show a progressive change toward shoshonitic compositions with time.

Trace Element Chemistry

General fractionation trends are displayed on 'Harker' variation diagrams of selected compatible and incompatible elements. Depletion of Sr with increasing degree of differentiation is consistent with the fractionation of plagioclase and apatite as indicated by the major element plots. The broader scatter of data points in some of the volcanic phases reflects the mobility of Sr during alteration processes. These samples also generally show the most variable K_2O and Rb contents. From 63 to 75 wt%, SiO_2, systematic increases in elemental abundances shown by Rb and Nb indicate that these elements behaved incompatibly. Extreme enrichment in these elements is shown by the most fractionated (> 76 wt"/SiO_2) granitoid samples.

A plot of Ti versus Zr is particularly useful for discriminating the various volcanic phases and plutonic pulses. Separate evolutionary paths are clearly defined with the least-evolved members of each group having distinct initial Zr/Ti ratios. Within each of the paired phases and pulses, both elements behave in a similar manner during fractional crystallisation. The steep trends in groups 1, 2, and 3 indicate that Ti is behaving more compatibly than Zr, whereas the shallower trends indicate that both elements have similar compatible behaviour. Differences in the rate of removal of these elements during fractionation are likely to be controlled by crystallisation of zircon and Ti magnetite. The available Nd and Sr isotope data rule out significant high-level crustal contamination. Differences in the initial Zr/Ti ratios for each group, therefore, are thought mainly to reflect differences in the composition of the source material.

The separate fractionation lineages displayed by each of the volcanic phases indicates a step-like evolutionary change towards Zr-enriched, Ti-depleted compositions with time. For example, Phase 3 compositions are less Ti-rich than Phase I but have higher Ti contents than Phase 2 at similar degrees of differentiation. Furthermore, there is marked compositional distinction between volcanic rocks of Phases 3 and 4 which were erupted during the same 143 Ma magmatic episode. This strongly suggests that magmas were being derived from two sources coevally. A similar, though less obvious distinction in source composition among coevally erupted magmas is shown for Phase 5 where two subphases can be distinguished: an early trachydacite-rhyolite subphase (SA), which is intermediate between Phases 3 and 4, and a late trachyte rhyolite subphase (5B), which is slightly less Ti-rich than Phase 4. The broad scatter of data points in Phases 1, 2, and 3 is partly due to subordinate evolutionary trends represented by individual formations.

The granitoids preserve patterns of Zr-Ti variation similar to those of the volcanic rocks although among the strongly differentiated compositions the fractionation lineages are less apparent. In general, Pulse 1 granodiorites and monzogranites, have very similar trends and abundances to volcanic rocks of Phase 1, although there is no volcanic equivalent for the Pulse lB monzogranites. Monzogranites and dykes of Pulse 2 also have similar Zr-Ti contents and fractionation trends to the volcanic rocks of Phase 2. Among the quartz monzonites, those of Pulse 3 are the most Ti-rich and may be more primitive members of the Pulse 3 monzogranite lineage. Comparing Pulse 3 granitoids with Phase 3 volcanic rocks, there is broad correspondence in both Zr Ti abundance and fractionation trend. Quartz monzonites of Pulse 4 have very similar Zr-Ti contents to Phase 4 volcanic rocks but lower Ti abundances compared to the coeval quartz monzonites of Pulse 3. Quartz monzonites of Pulse 5 can be split into an earlier pulse (SA) and a later pulse (5B) which broadly correspond with the volcanic subphases (SA and 5B).

Whereas quartz monzonites of Pulse SA lie on a fractionation trend that would give rise to the monzogranites, those of Pulse 5B are distinctly less Ti-rich, and lie on a fractionation lineage corresponding to a small group of slightly younger, less Ti-rich monzogranites. Although no volcanic equivalents have yet been identified for Pulse 6, these rocks record the lowest Ti and highest Zr contents of any magmatic group in Hong Kong. There is a possible fractionation lineage linking the quartz monzonites and monzogranites of this youngest recognised pulse. On MORB-normalised spider diagrams, representative compositions from the five phases of volcanic rocks have consistent profiles that are similar to those magmas found in active continental margins. In particular, the profiles show enrichment of K, Rb, and Th, and depletion of Nb, which are characteristic features of subduction-related magmas and an indication of crustal involvement in magma genesis. Troughs for Ba and Sr can be explained by fractional crystallisation of feldspar whereas the negative Ti anomaly shown by all the analysed compositions reflects the crystallisation of Ti-magnetite.

MORB-normalised trace element patterns for representative granitoids from each pulse are similar to those of the volcanic rocks and a similar relationship is exhibited by Phase 2 and Pulse 2 rocks on a chondrite-normalised plot of the REE data. The chondrite-normalised plot also highlights the similarities in REE behaviour among the Middle Jurassic Lower Cretaceous granitoids. In general, from Pulse I through to Pulse 5, REE abundances increase while preserving the light REE-enriched shape of pattern. Sewell et ail (1992) have drawn attention to the rather flat to slightly heavy REE enriched patterns shown by the A-type granitoids of Hong Kong which have overall REE concentrations of up to sixty times chondrite abundance. The quartz monzonites commonly show a slight negative dip in the middle REE pattern which is thought to be caused by amphibole fractionation. The Middle Triassic Deep Bay Granite has markedly lower heavy REE contents compared to the Jurassic-Cretaceous granitoids ruling out a genetic relationship.

The general pattern of evolution among the Mid-Jurassic Early Cretaceous magmas is for the least-evolved compositions of each group to become more K_2O enriched, Zr-enriched and Ti-depleted with time. Separate batches of magma generated at 164 159 Ma, 146 Ma, 143 Ma, and 140 Ma were derived from variable degrees of partial melting of crustal material involving a mantle-derived component. Each batch of magma was characterised by a slightly different initial Zr/Ti ratio and differentiated mainly by fractional crystallisation involving the removal of plagioclase, K-feldspar, biotite, amphibole and Ti-magnetite.

Darbyshire & Sewell have isotopically modelled the sources of Hong Kong granitoid magmas in terms of a mantle component and two crustal components. Based on Sr and Nd isotope data, they propose that the source of the Lamma Suite granitoids (Pulse 1) in the northwest of Hong Kong is a mixture between a juvenile mantle source and dominantly Archaean crustal material whereas in the southeast of Hong Kong, the source of the Lion Rock Suite granitoids is a mixture between a mantle-derived melt and dominantly Proterozoic crustal material. They interpret the variable isotope signatures in the area between as marking the boundary between these two dominant crustal sources and recording the strongest mantle-derived imprint.

The overall consistency of REE patterns and abundances among the I-type granitoids of Hong Kong is in marked contrast to the differences in the Nd isotope initial ratios. One explanation for this disparity could be that the isotope signatures are more sensitive indicators of variations in the crustal source component. This situation could arise if the isotope signatures of the crustal component did not have sufficient time to re-equilibrate following assimilation by juvenile mantle-derived melts. Alternatively, the REE patterns and abundances of the I-type granitoids may largely reflect those of the mantle-derived component. The strongly peraluminous composition and negative ENd signature (-13.9) of the Middle Triassic Deep Bay Granite preclude an appreciable mantle-derived component in the source region. In this instance, the distinctive REE pattern of the Deep Bay Granite is considered to reflect that of the dominant Archaean crustal source component. If the REE contents of the I-type granitoids were also primarily related to those of the crustal component, then their REE patterns would be expected to vary in concert with their isotope signatures. Since this is not the case, the consistent REE patterns and abundances displayed by the Upper Mesozoic rocks are thought to largely reflect those of the mantle-derived component.

The involvement of both Archaean and Proterozoic crust in the genesis of Hong Kong magmas has largely been confirmed by the presence of inherited zircons in the granitoids and related volcanic rocks. Together with the REE evidence for a consistent mantle-derived signature, we propose that the variability in magmatic composition is best explained by interaction of a mantle-derived melt with differing proportions of Archaean and/or Proterozoic crustal material.

The higher $^{87}Sr/^{86}Sr$ initial ratios and lower õ Nd signatures of Pulse I granitoids compared with the other groups suggest that mantle-derived melts interacted strongly with an Archaean-dominated crustal source. A strong crustal and mantle influence is also preserved in the source area of Pulses 4 and 5 although the isotope data indicate that the crustal contaminant is likely to have been dominantly Proterozoic in age. The strongest mantle influence is preserved in the sources of Pulse 2 and Pulse 3 granitoids. Geographically, both groups lie within the Zone 2 of Darbyshire & Sewell which has been interpreted as representing a region of deep crustal faulting. These discontinuities of crustal dimensions may have promoted the mixing of mantle-derived magmas with both Archaean and Proterozoic crustal sources.

Campbell & Sewell have shown that there are fundamental structural controls on the distribution and disposition of volcanic centres in Hong Kong which are typical of volcanism within the Lianhuashan Fault Zone. They conclude that the onset of transtension and extension in the Mid- to Late Jurassic and Early Cretaceous led to the development of large calderas and high-level subvolcanic magma reservoirs. Lithologically, the volcanic rocks of Phase 3 are mainly composed of coarse ash crystal tuff of dacite to rhyolite composition. The crystal-rich character of these deposits suggests the development of high-level, subvolcanic magma chambers which permitted prolonged crystal fractionation prior to eruption. These magma chambers are now preserved as large elliptical plutons (Pulse 3) which have demonstrable links to the overlying volcanic rocks. By contrast, most of the volcanic rocks belonging to Phase 4 are trachyte to rhyolite in composition and their plutonic equivalents occur mostly as quartz monzonite dykes. The volcanic deposits of Phase 4 are dominantly composed of fine ash vitric tuff, commonly thick and densely welded indicating that they were emplaced rapidly and at high temperature. These deposits mainly accumulated adjacent to linear fissure vents. Thus, two main types of volcanic plutonic assemblage can be recognised: 'caldera-type' in which volcanic deposits accumulated in tectonomagmatic depressions controlled mainly by ellipsoidal and rectilinear fracture systems, and 'fissure-type' in which volcanic deposits accumulated adjacent to fissure vents controlled by linear fracture systems. The two best examples of 'fissure-type' plutonic-volcanic assemblages include the Phase SA assemblage and the Phase SB assemblage, whereas the two best examples of caldera-type' assemblages include the Phase 3 assemblage and the Phase lA assemblage. It is, therefore, conceivable that the low Zr/Ti, trachyte-rhyolite 'fissure-type' magmas migrated relatively rapidly to the surface along deep crustal pathways and did not develop large subvolcanic reservoirs.

Caldera development, coupled with a transcurrent movement, is a feature of some back-arc continental settings. Campbell & Sewell postulate that the emplacement style and character of the Phase I volcanic rocks and related Pulse 1 intrusions suggest the incipient development of a back-arc setting in which extension and transtension are the dominant tectonic controls. Such a setting is likely to promote the development of high-level, magma storage reservoirs and associated calderarelated volcanic activity. Extension and transtension can also be regarded as the main tectonic controls on the development of 'caldera-type' volcanic-plutonic assemblages in Hong Kong, especially where relatively rapid extension can be demonstrated. Such rapid extension may have promoted the passage of large volumes of strongly mantle-influenced magmas to the surface.

On the other hand, transtension may have promoted the interaction of mantle-derived magmas with continental crust resulting in the generation of increasingly K_2O-rich and TiO,-depleted magmas. These magmas then rose relatively rapidly to the surface through transtensional crustal pathways to be extruded predominantly along linear crustal fissures.

Precise dating of the Mesozoic volcanic and plutonic rocks of Hong Kong has placed very tight controls on the sequence of magmatic events. Combined with stratigraphic, isotope and geochemical data, it has been possible to demonstrate the existence of co-magmatic relationships among coeval volcanic and plutonic pairs indicating an inextricable link between eruptive products and their subvolcanic sources. The earliest calc-alkaline andesitic eruptives are likely to represent a remnant of an extensive andesitic volcanic arc province that dominated the Eurasian plate margin during the Early to Mid-Jurassic. Although these arc-type rocks have not been precisely dated, the marked change in composition between these and the Middle Jurassic Phase 1 rocks suggests that either the intervening rocks have been eroded away, or that there was a period of relative volcanic quiescence in this part of the southeast China igneous province.

Renewed volcanic activity of compositionally different character suggests that the westward-dipping subduction zone may have migrated farther east producing high-K calc-alkaline compositions distinctive of a mature continental arc province.

This period of Mid-Jurassic to Early Cretaceous magmatism was punctuated by short bursts of activity during which the overall tectonic setting changed from mature continental arc to back-arc. Periods of rapid extension-dominated tectonism promoted the development of calderas, dyke injection, and high-level subvolcanic magma reservoirs. Periods of transtension promoted the magmatic activity along linear fissures. The increasing K20-rich character of the magmatic products possibly reflects further eastward migration of the subduction zone as the depth to the Benioff zone increased. Shoshonitic compositions represent the final products of the Mid-Jurassic-Early Cretaceous magmatism prior to sudden termination at 140 Ma.

Geochemistry of Pelagic and Hemipelagic Carbonates

Recent work on Upper Cretaceous sequence stratigraphy in European basins has demonstrated that chalks provide valuable information on short- and long-term sea-level change. Sedimentary relationships on basin margins, combined with stratigraphic variation in lithology and carbon isotopes, have been used to define sequence boundaries, onlap surfaces and systems tracts. However, work on other Mesozoic sequences suggests that the elemental composition of pelagic sediments also varies in response to sea-level change.

Perhaps the best documented eustatic rise in sea level in Earth history occurred during the Cenomanian Stage with the global expansion of epicontinental seas and the widespread establishment of pelagic and hemipelagic carbonate deposition. The Cenomanian stratotype was used by Haq et aL (1988) to illustrate sequence stratigraphic analysis of an outcrop section, and a detailed sequence stratigraphic framework has been proposed recently for the Cenomanian of the Anglo-Paris Basin by Robaszynski et aL (1998). In this chapter we present the first systematic study of the elemental geochemistry of a Cenomanian Chalk succession. We test relationships between elemental and carbon isotope chemostratigraphy and the previously established sequence stratigraphy, and demonstrate that geochemical criteria provide a valuable tool for the recognition of sea-level change.

Culver Cliff on the eastern Isle of Wight, southern England, provides complete exposure of Cenomanian to Campanian chalks, traditionally assigned to the Lower, Middle and Upper Chalk formations. The section offers fresh exposures of pale green (unoxidised) sediments which have suffered only modest burial diagenesis, making them particularly well-suited to geochemical studies. The Cenomanian-Turonian (Lower and basal Middle Chalk) section was described by Kennedy (1969) and parts were studied more recently by Gale and Paul et al. (1994), who correlated these with various localities in NW Europe. The entire section was logged by us in detail and key litho- and biostratigraphic marker beds identified. As with other northern European sections, a hiatus representing the uppermost Albian-lowest Cenomanian occurs at the base of the Chalk (Upper Greenland-

Glauconitic Marl boundary), but the Cenomanian-Turonian succession at Culver compares favourably with those exposed elsewhere in the region.

Sequence Stratigraphic Framework

The Cenomanian of NW Europe typically consists of chalks and marls, rhythmically bedded on a decimetre-scale. This rhythmicity has been attributed to Milankovitch climatic forcing, probably driven by the 20 ka precession cycle, modulated by the short eccentricity cycle (100 ka bundles). However, lower frequency, tens of metres thick cycles are also recognised. Key horizons in these cycles have been correlated to onlap surfaces on the basin margin and locally on Palaeozoic massifs, providing a basis for placing these low-frequency cycles in a sequence stratigraphic framework. The Cenomanian is particularly well suited to sequence stratigraphic analysis because biostratigraphic resolution based principally on ammonites, is very high, enabling precise correlation of strata on a basinal scale.

Based on stratal geometry and variation in elastic input, Robaszynski et al. (1998) proposed six depositional sequences for the Cenomanian, each in the order of 1 Ma duration. Systems tracts were characterised and defined by diverse criteria including the geometry of sedimentary bodies, basin margin onlap, discontinuity surfaces (omission surfaces and hardgrounds) and lithological characters, particularly variations in quartz sand content, clay: carbonate ratios, and the presence of authigenic minerals, glauconite and francolite.

Eight major discontinuity surfaces that have been traced throughout the Anglo-Paris Basin, combined with other key marker horizons recognised in East Kent and the Boulonnais, may be used to correlate Cenomanian sections throughout southern England. These discontinuity surfaces and marker beds were used to place the sequence boundaries (SB), transgressive surfaces (TS) and maximum flooding surfaces (MFS) defined by Robaszynski et al. (1998) in the Culver Cliff succession. These surfaces define the positions of lowstand, transgressive, and highstand systems tracts which were assigned sequence numbers according to the scheme of Robaszynski et al. (1998).

Sequence I The basal Glauconitic Mart contact with the Gault Clay in East Kent and the Upper Greensand on the Isle of Wight, represents low (but not lowest) Cenomanian Neostlingoceras carcitanense Subzone, Mantelliceras mantelli Zone, resting disconformably on high Albian Stoliczkaia dispar Zone. The contact

(DSI) is coincident with the sequence boundary/transgressive surface at the base of Sequence 1. The maximum flooding surface of Sequence 1 is located in the basal Chalk Marl below a series of beds containing abundant Inoceramus crippsii Mantell bivalves. These beds occur < 1 m above the top of the Glauconitic Marl at Culver.

Sequence 2 Nodular limestone M3 of Gale (1989a) contains common Schloenbachia varians (J. Sowerby) and other ammonites typical of the uppermost N. carcitanense Subzone. The top of the bed is a prominent omission surface (DSII) overlain by a dark mart piped downwards by large Thalassinoides burrows. The omission surface marks the boundary between sequences 1 and 2, and represents a sequence boundary overlain directly by a transgressive surface.

The maximum flooding surface of Sequence 2 is not clearly defined in East Kent. At Culver we place it in the lower Chalk Marl, within a series of beds containing abundant Exanthesis labrosus (T. Smith) sponges.

Sequence 3 Phosphatic dark silty marl M4 lies within the Mantelliceras saxbii Subzone of the M. mantelli Zone, I-2 rn below the base of the Mantelliceras dixoni Zone. The discontinuity at the base of this marl (DSIII) marks the transgressive surface of Sequence 3, with the sequence boundary occurring below, in the immediately underlying many interval. The two surfaces may be amalgamated at Culver, or they may be so closely spaced that they were not resolved by our sampling.

The maximum flooding surface of Sequence 3 lies in the lower part of a distinctive sequence of less many chalk couplets in the middle M. dixoni Zone. Sequence 4 Orbirhynchia mantelliana band 2, named after the common occurrence of the brachiopod Orbirhynchia mantelliana (J. de C. Sowerby), is located at base of the Middle Cenomanian Cunningtoniceras inerme Zone. The sequence boundary of Sequence 4 lies at the base of the marly interval that characterises 0. mantelliana band 2.

Transgressive surface 4 is placed at the base of the Cast Bed of Price (1877), couplet Cl of Gale (1995), a dark silty marl which at the base contains an abundance of the small scallop Entolium orbiculare (J. Sowerby) and a diverse fauna including the belemnite Actinocamax primus Arkhangelsky. The maximum flooding surface of Sequence 4 is the calcimetry break at the base of couplet C 17, which marks the base of the Grey Chalk.

Sequence 5 A distinctive many chalk overlying an omission surface (DSV) coincident with the base of the Acanthoceras jukesbrownei Zone defines the sequence boundary of Sequence 5. Approximately 2 m above this, a distinctive thin grey marl (base of couplet DI) marks the bottom of a sequence of silty and arenitic chalks containing numerous laminated structures that have been variously described as burrow fills or scours. This interval (couplets 131-138) is commonly referred to as Jukes-Browne Bed 7, the base of which constitutes the transgressive surface of Sequence 5. Pycnodonte vesiculare (Lamarck) oysters are common above this surface.

The maximum flooding surface of Sequence 5 is located close to the Middle-Upper Cenomanian boundary (=base of the Calycoceras guerangeri Zone).

Sequence 6 The base of the Plenus Marl (DS VI; sub-plenus erosion surface) marks the sequence boundary for Sequence 6. The base of Plenus Marl Bed 4 is the transgressive surface (DSVII) of Sequence 6.

The base the Ballard Cliff Chalk (Melbourn Rock) constitutes a major facies change from many to nodular and intraclastic chalks (DSVIII). The maximum flooding surface of Sequence 6 is placed at the base of the overlying Holywell Chalk (=Meads Marl 6 of Mortimore 1986).

The above sequence stratigraphy formed the basis for comparison with our new chemostratigraphic data. However, in the North Cantabrian and Basco-Cantabrian basins of northern Spain, an additional sequence of earliest Late Cenomanian age is recorded, the base of which falls at or near the base of the Eucalycoceras pentagonuml Calycoceras naviculare (=C. guerangeri of this study) Zone.

Analytical Methods

Bulk sediment (125) samples were collected at c. I m intervals. Chalks and marls were both sampled; no attempt was made to select particular lithologies. Sample spacing was reduced across critical stratigraphic intervals such as the base of the Middle Cenomanian and the Cenomanian-Turonian boundary because these have been interpreted previously as representing periods of rapid environmental change which should be reflected in the geochemical signatures. Samples were ground to fine powders in an agate ball mill, washed with deionised water to remove modern sea-salt contamination and dried at 60 deg C; subsamples were fused with LiBO, and the melts dissolved in 0.5 M HN03. Silica, Ti, Al, Mn, Ca and Zr were determined by inductively coupled plasma-atomic emission spectrometry (ICP-AES) at Kingston University. Analytical procedures and numerical data are presented elsewhere.

A complementary suite of 85 samples was collected for stable isotope analysis. Samples were cleaned using 100, 1-12_02 followed by acetone, and dried at 60 deg C. Powders were reacted with purified orthophosphoric acid at 90 deg C and analysed on-line using a VG Isocarb device and PRISM mass spectrometre at the University of Oxford. Data are reported relative to the VPDB (Vienna Pee Dee Belemnite) standard; further analytical details are provided by Jenkyns et al (1994). Data for an additional 20 mid-Cenomanian samples were taken from Paul et al (1994).

Previous workers have used acetic acid leaches in an attempt to isolate 'carbonate-associated' elements in pelagic carbonates. However, these leaches dissolve both carbonates and manganese oxyhydroxides, and cause desorption of Mn from mineral surfaces. Experiments by Murphy (1998) have shown that 1 M acetic acid dissolves 90-99% of the Mn in Cenomanian chalks, and whole-rock trends are essentially identical to those obtained from partial leaches. Manganese shows no clear correlation with either Ca or Al, and is plotted here as elemental and Mn/AI curves; Mn/Ca essentially follows Mn. Whole-rock Ca data are presented as $CaCO_3$ since low-Mg calcite is the only significant Ca-bearing phase. Silica, Ti and Zr covary with Al due to carbonate dilution effects, but when normalised to Al provide valuable information on mineralogical variation in the siliciclastic fraction.

Manganese Chemostratigraphy

The Mn profile for Culver Cliff displays a general long-term decrease from the base of the Cenomanian to the Turonian. This corresponds to declining Si, Ti, Al, Zr and bulk insoluble residue contents that indicate a declining siliciclastic supply during the Cenomanian transgression. This is consistent with Mn being controlled partly by the detrital flux. However, Mn is assumed to be predominantly a carbonate-bound element, so the partitioning of Mn in chalks must be at least a two-component system with Mn being associated with both clay minerals and carbonate (with secondary Mn-oxyhydroxides also being developed in weathered sections). Superimposed on the long-term trend are seven short-term cycles with prominent Mn maxima that do not correlate with either Al or Ca. The maxima of most of these cycles occur around the maximum flooding surfaces defined by Robaszynski et aL (1998), while minima are close to, or coincident with, sequence boundaries. Lowstand systems tracts generally have relatively low Mn contents, transgressive systems tracts display sharply rising Mn values that reach maxima around maximum flooding surfaces, before declining again through highstand systems tracts.

In addition to the six Mn cycles corresponding to the six established sequences in the Anglo-Paris Basin, a further cycle occurs in the Upper Cenomanian. Based on the Mn stratigraphy at Culver, the sequence boundary for this sequence is placed at the Mn minimum in the Amphidonte bed, a marker interval in the C guerangeri Zone containing two or three marls and characterised by the presence of occasional Amphidonte and subordinate Pycnodonte oysters.

The Amphidonte bed is a basin-wide marker that can also be traced throughout NW Germany. The transgressive surface of Sequence Sa is placed at a level of sharply rising Mn values occurring approximately 4 m above the Amphidonte bed, in a prominent thin grey flaser marl seam that overlies an omission surface. The maximum flooding surface is placed at the Mn maximum on the smoothed profile, which occurs at the base of sequence of three very prominent 10 cm thick dark marl seams in the middle C guerangeri Zone, termed here the Monument Marls. The lowest of these (Monument Marl 1) has a clearly defined omission surface at its base.

The long-term decrease in Mn is eliminated in the Mn/Al ratio profile, but the trends are much less coherent. Nonetheless, major peaks and troughs identified on the Mn profile are also displayed by Mn/Al, with the exception of the large Mn peak associated with the Cenomanian-Turonian boundary. Here, the sharp rise in Mn at the base of the Plenus Marl is eliminated, and the Mn maximum is shifted into the Lower Turonian, so long-term trends in Mn/Al ratios appear to reflect the sequence stratigraphy better.

Carbon Stable-isotope Stratigraphy

Calcimetry and Variation in the Siliciclastic Fraction

The calcimetry profile, shows reasonable agreement with the sequence stratigraphic model that predicts low carbonate in lowstand systems tracts, rising values from transgressive surfaces through transgressive systems tracts, with maxima at maximum flooding surfaces that plateau through the overlying highstand systems tracts. This correspondence is to be expected because clay: carbonate ratios were one of the criteria used by Robaszynski et al. (1998) to define their sequence stratigraphic framework in basinal sequences. Deviations from the predicted relationships may be explained in part by the relatively course sampling resolution adopted in the chemostratigraphic study. As expected, the Al profile is close to a mirror image of Ca and corresponds well to the gamma-ray logs for Culver and a closely correlative succession in the Boulonnais, northern France.

Transgressive surfaces and maximum flooding surfaces are picked out by positive excursions in Ti/Al, whereas sequence boundaries generally exhibit low Ti/Al contents; Zr/AI shows similar trends. It is notable that in most cases peaks are defined by several samples (i.e. extend over more than I m) and consequently represent relatively long-term (perhaps 60-100 ka, based on the cyclostratigraphy) changes in sediment composition, rather than reflecting mineralogical and/or chemical variation associated with sediments overlying individual omission or erosion surfaces. Titanium resides principally in heavy minerals: ilmenite, rutile, anatase and sphene, whereas Zr is located predominantly in zircon. Where a basaltic volcanogenic source can be excluded, increased Ti/Al ratios in deep-sea sediments correlate with increased current energy and detrital supply. Chalk tephras have low Ti contents that contrast to elevated Ti in the detrital fraction, so high Ti/Al ratios in chalks are interpreted as indicating pulses of increased current energy and siliciclastic supply.

Silica occurs in a much wider range of sediment constituents than Ti and Zr, including detrital quartz, biogenic opal (and its diagenetic products: opal-CT, chalcedony, quartz), glauconite and clay minerals. Interpretation of the Si/Al profile is ambiguous, but high and declining Si/Al values from the base of the Cenomanian and high Si/Al ratios at transgressive surfaces and maximum flooding surfaces correlate with increased detrital quartz and/or glauconite contents. The correlation between Ti and Si peaks is consistent with at least some Ti residing in titanomagnetite inclusions in detrital quartz, like those identified from Coniacian-Campanian chalks, also on the Isle of Wight.

loA

Manganese Chemostratigraphy

General trends in the Mn chemostratigraphy show remarkably consistent trends with respect to the sequence stratigraphic model.

One discrepancy relates to the location of the Sequence 4 sequence boundary which is apparently overlain by sediments with high (but falling) rather than low Mn values. However, clear Mn and calcimetry minima on the smoothed profiles both occur in the Cast Bed, the base of which was interpreted by Robaszynski et aL (1998) as representing a transgressive surface. Repositioning the sequence boundary of Sequence 4 at the base of the Cast Bed, an alternative favoured by some workers, would remove this discrepancy. However, this would place the Ti, Zr and Si maxima at the sequence boundary rather than at the transgressive surface. The possible relocation of this sequence boundary warrants further study. A second apparent anomaly relates to the lowstand systems tract of Sequence 6 which has a high Mn content. However, Mn/Al ratios remain low, indicating that high Mn values are probably caused predominantly by the marked increase in clay content that occurs at the base of the Plenus Marl.

The Mn stratigraphy at Culver records the existence of an early Late Cenomanian sequence that corresponds in biostratigraphical position (lower C. guerangeri Zone or its equivalents) with that described from northern Spain by Wilmsen (1997) and Grafe & Weidmann (1998). Robaszynski et al. (1998) did not find any sedimentological evidence for this sequence in the Anglo-Paris Basin, but Fayau Hardgrounds 1 and 2 developed in shallow-water sands in Sarthe, NW France are here interpreted as expressions of, respectively, (1) the sequence boundary and (2) the transgressive surface of this sequence. The chalk facies was probably deposited in water too deep to clearly reflect the relatively small sea-level changes responsible for the formation of the C. guerangeri Zone sequence.

Marine Geochemistry of Mn

Manganese and Carbon-isotope Stratigraphy

Increased mid-ocean ridge volume caused by a period of enhanced spreading is regarded as being the main cause of eustatic rise in sea level during the Late Cretaceous. Crustal production rates during the Cenomanian Turonian were at their highest for the last 100 Ma, and were associated with increased mid-ocean ridge volcanism and the formation of extensive ocean basaltic plateaux. The decline in seawater 17 Sr/"Sr ratios through the Cenomanian supports a substantial hydrothermal input into the oceans, but the maximum rate of sea-level change driven by this mechanism is assumed to be in the order of 10 m Ma-.

The six depositional sequences in the Cenomanian each represent c. I Ma, so changes in sea-floor spreading rate would have to occur more rapidly than previously thought to explain such short term sea-level fluctuation. There is no evidence of permanent polar ice caps during the mid Cretaceous making a glacio-eustatic mechanism for rapid sea-level change unlikely, although episodes of high-latitude ice accumulation and associated sea-level fall have been postulated. Manganese enriched waters and bottom sediments are produced by hydrothermal processes at mid-ocean ridges, so extensive ridge volcanism and ocean plateau formation during the Cenomanian and early Turonian may have increased the global marine Mn flux, while pulses in hydrothermal activity may have been partly responsible for the rising Mn contents that characterise Cenomanian transgressive systems tracts.

Manganese deposition reached a minimum in the early Turonian during the overall peak Cretaceous highstand. It may be speculated that low Mn also relates to elevated atmospheric and oceanic P02 associated with low pCO_2 values that followed the Cenomanian Turonian boundary event. Higher oxygen concentrations in marine waters would have led to increased oxidation of organic matter in the water column and deeper and weaker sediment redox boundaries, both of which would have decreased the effectiveness of Mn redox cycling and Mn-carbonate precipitation.

Idealised relationships between chemostratigraphy, depositional sequences and sea level are summarised. Manganese exhibits minima around sequence boundaries and through lowstands, rising values from the transgressive surfaces through transgressive sequence tracts, maxima around maximum flooding surfaces, and declining values through highstands. Remarkably similar relationships between Mn curves and sequence stratigraphy have been described in Toarcian-Oxfordian, Tithonian Berriasian and Coniacian Maastrichtian pelagic limestones elsewhere, but previous authors have not examined the behaviour of other elements. Carbonate contents increase in response to rising sea level, but display sharp decreases at sequence boundaries and minima in lowstands systems tracts. Values increase through each transgressive systems tract, reach a maximum around the maximum flooding surface, and then remain high and constant through the highstand systems tract. Silica and trace-element (Ti, Zr) aluminium ratios peak around transgressive surfaces and maximum flooding surfaces, indicating pulses of increased siliciclastic input which are not confined to single omission/ erosion surfaces. This study has demonstrated that when Mn is combined with calcimetry, Si, Ti and Zr data, elemental chemostratigraphy provides robust criteria for developing sequence stratigraphic models and identifying sea-level change in pelagic-hemipelagic carbonate successions. However, manganese enrichment in Cretaceous carbonate sediments does not relate directly to oceanic anoxia in these sequences, so models for the formation of Mn ore deposits that invoke such a relationship may require re-evaluation.

Isotopic Constraints on Subduction Polarity

The Kohistan terrane of northern Pakistan is a superb example of juvenile crust that formed by magmatic additions at an intra-oceanic convergent margin. It is particularly important to tectonicists because its formation and accretion history constrains the Cretaceous palaeogeography of the Neotethys, and to igneous geochemists because it provides a complete and well-exposed section of primitive arc crust. Almost no isotopic data has been previously reported for the Kohistan intraoceanic arc terrane, and several important aspects about its tectonic significance remain controversial, particularly how wide was the oceanic seaway that separated it from Asia, and what was Kohistan's subduction polarity? We report here new major and trace element and isotopic compositions for igneous rocks that formed during the evolution of Kohistan as an intra-oceanic arc, and use these data to (1) confirm that Kohistan represents an intra-oceanic arc; (2) demonstrate that, prior to collision with the Karakorum terrane, the Kohistan arc faced north over a south-dipping subduction zone; and (3) infer that the Kohistan intra-oceanic arc originated near the present location of the DUPAL isotopic anomaly, at near equatorial latitudes.

Geological Setting

The Kohistan terrane is located in northern Pakistan and is bounded to the north by the Shyok (or Northern) Suture and to the south by the Indus Suture. Sedimentary sequences indicate that formation of the intra-oceanic Kohistan arc began in Early Cretaceous time, consistent with Ar/Ar and Rb/Sr ages indicating that the bulk of the igneous infrastructure formed between 110 and 90 Ma. The inference that Kohistan originated as an intraoceanic are results from its mafic bulk composition. presence of pillowed lavas and marine sediments, low-K nature of associated felsic rocks, absence of continental crust or detritus, and separation of the inferred intra-oceanic arc of Kohistan from continental crust to the north and south by unequivocal suture zones. Kohistan and Ladakh comprise correlative parts of this Cretaceous intraoceanic arc separated by the late Cenozoic Nanga Parbat Haramosh uplift. The intraoceanic phase of Kohistan lasted until sometime between 102 and 85 Ma, when Kohistan collided with the Karakorum (Asia) across the Shyok Suture. From this time until collision with India about 50 Ma ago, Kohistan existed as an Andean-type margin. This report focuses on the intra-oceanic phase of Kohistan.

The intra-oceanic arc crust of Kohistan consists of five principal units, extending from the Indus Suture in the south to the Shyok Suture in the north: (1) basal ultramafic mafic cumulates (Jijal Ultramafics); (2) Kamila Amphibolite; (3) Chilas Complex of mafic to intermediate plutonic rocks; (4) early bimodal suite of intrusive rocks and Gilgit gneisses; and (5) Chalt Volcanics. The early bimodal suite is a small part of the region. The bulk of the Kohistan batholith formed after collision of Kohistan and Karakorum, during the Andeanmargin phase of Kohistan, and is of no concern to this report. We report geochemical and isotopic data on Kamila amphibolites, Chilas Complex intrusive rocks, and Chalt Volcanics.

The Kamila Amphibolite belt is a composite mass dominated by amphibolite-facies meta-plutonic and metavolcanic rocks. Hornblende Ar Ar cooling ages of about 80 Ma indicate that regional amphibolite facies conditions existed prior to this time. Two types of amphibolites are identified in the field: (1) fine- to medium grained amphibolites, either homogeneous or banded and (2) homogeneous, medium-to coarse-grained amphibolites. The former are thought to be metamorphosed mafic to intermediate volcanics, whereas the latter are interpreted as metamorphosed gabbros and diorites. Our study was concerned with the metavolcanic amphibolites; we report no data for the second group. The Kamila metavolcanic amphibolites are subdivided into HFSE-enriched and -depleted groups, and are referred to as 'E-type' and iD-type` metavolcanic amphibolites by Khan et al. (1993). E-type amphibolites have MORB-like major and trace element compositions and do not show spikes in normalised trace element patterns characteristic of subduction-related igneous rocks. In addition they show REE patterns with modest depletions in LREE. In contrast, D-type amphibolites show normalised trace element patterns consistent with the participation of a `subduction component' (Pearce 1984). Kamila amphibolites derived from metamorphosed plutonic rocks show a very strong `subduction component', with high Ba/Nb. Kamila metaplutonic amphibolites are compositionally very similar to gabbronorites of the Chilas Complex. Khan et al. (1993) concluded that all three varieties of Kamila amphibolites were derived from a mantle source that was enriched in incompatible trace elements.

The Chilas Complex is more than 300 km long and up to 40 km wide, defining the spine of the Kohistan terrane. The Chilas Complex is subdivided into 'main gabbronorites' and a 'UMA association' by Khan et al. (1989). Gabbronorites comprise the vast bulk of the Chilas Complex; the distinctive cumulate ultramafic, mafic, and anorthositic rocks of the UMA association formed later. Gabbronorites define a calc-alkaline suite that range from about 51 to 59o/0 SiO_2, comparable to island-arc non-cumulates of Beard (1986). They show the distinctive chemical characteristics of convergent margin magmas, especially strongly positive spikes for Ba, K, and Sr, and negative Nb anomalies. They have modest LREE enrichments $((Ce/Yb)_n$=2-3.5), with REE contents of about 10 x chondritic abundances. The rocks cooled and equilibrated under P-T conditions of the pyroxene granulite facies, estimated at 750-850 deg C and 5 6.5 kbar. In spite of the close temporal and spatial affinities of the two principal Chilas Complex components, significant differences in mineral compositions exist between gabbronorites (Olivine absent, magnetite-ilmenite, An_{40}- An_{64}) and UMA (Olivine abundant, Cr-spinel, An_{83}- An_{99}). The Chilas Complex was emplaced at c. 100 Ma into high levels of the deformed arc. For this report, we accept the Chilas Complex as developed during the intra-oceanic phase of Kohistan arc evolution, although Treloar et al. (1996) conclude that the Chilas Complex was emplaced after Kohistan collided with the Karakorum.

The Chalt Volcanics occupy a 330 km x 30 km belt along northernmost Kohistan. The upper parts of these volcanics are interbedded with the Yasin sediments of Aptian Albian age. Abundant pillowed lavas demonstrate that some if not all of these lavas erupted subaqueously. The Yasin and Chalt are strongly deformed and modestly metamorphosed, from lower greenschist facies in the west to lower amphibolite facies in the east. A significant compositional difference is observed along the strike of the belt, with a high-Mg suite encountered in the Hunza Valley (where the present suite was obtained) and a low to intermediate Mg calc-alkaline sequence found to the west. Boninite is common in the Hunza Valley section along with some tholeiitic basalts and abundant felsic lavas.

Analytical Techniques

Major- and trace-element analyses were obtained using standard XRF spectrometry techniques. The samples of Kamila Amphibolite and the Chilas Complex were analysed at the University of Oklahoma, using a Rigaku XRF, while those from the Chalt Volcanics were analysed at the University of Leicester, using Philips PW 1450 automatic XRF. Similar analytical techniques and standards were used at the two laboratories. Major elements were determined from glass fusion beads (made using a lithium tetraborate/ lithium metaborate flux; rock:flux ratio 1:5) analysed by using a Rh anode X-ray tube. Trace element concentrations were determined on pressed-powder pellets made using an organic binder (Moviol). Ni, Zr, Nb, Rb, Sr, Y determinations were made using a Rh anode X-ray tube with mass absorption corrections applied using the intensities of the Rh K alpha Compton scatter peak. Determinations of Cr and Ba used a W anode X-ray tube with mass absorption corrections applied using the intensities of the W L alpha Rayleigh scatter peak and Fe Ku to cross the Fe-absorption edge after the technique of Nesbitt et al. (1976). Detections limits (2Ó) are approximately I ppm for Cr, Rb, Sr, Zr, Nb and Y, and 2 ppm for Ni, Cr and Ba.

Concentrations of Nd, Sm. and Pb were determined by isotope dilution at UTD. Standard isotope dilution and cation exchange techniques were used to separate Sr. and LREE (Nd & Sm). A procedure modified after Richard et al. (1976) was used to separate Nd from Sm for isotope composition runs. The Pb separation procedure is reported in Manton (1988), with an initial anionic column separation. Total processing blanks were: < 140 pg Pb; <700 pg Nd and < 1 ng Sr. Isotopic compositions were determined with a Finnigan MAT 261 multi-collector thermal ionisation mass spectrometre; Nd was analysed in the dynamic multicollection mode whereas Sr and Pb were analysed in the static multicollection mode. Sr isotopic compositions were fractionation-corrected to $^{86}Sr/^{88}Sr$=0.1194 and adjusted relative to E&A SrCO3 $^{87}Sr/^{86}Sr$=0.70800. Nd isotopic compositions were fractionation-corrected to $^{146}Nd/^{144}Nd$=0.7219 and ENd was calculated using the aNd values of Pier et ul (1989) for BCR-1 and UCSD standards. The standard values with reproducibility during the year in which isotopic analyses were carried out (I standard deviation, with digits corresponding to the last decimal places reported) were: E&A SrCO3 $^{87}Sr/^{86}Sr$=0.708020 (35; n=32), UCSD $^{143}Nd/^{144}Nd$=0.511848 (14; n=26), and BCR-1 $^{143}Nd/^{144}Nd$=0.512620 (8; n=8); NBS 981 (with

0.15% per amu Pb fractionation correction) =”*Pb/2')4Pb= 16.943, $^{207}Pb/^{204}Pb$=15.495. $^{208}Pb/^{204}Pb$=36.736 for 16 analyses.

Sr and Nd isotopic compositions listed are reported both as measured following fractionation correction ($^{143}Nd/^{144}Nd$) and, in the case of Sr, adjustment to a value of 0.70800 for E&A SrCO3 $^{87}Sr/^{86}Sr$, as well as initial isotopic compositions corrected for 120 Ma of radiogenic growth using values of $^{147}Sm/^{144}Nd$ and $^{87}Rb/^{86}Sr$ from element abundances listed. Initial isotopic compositions are reported either as $^{87}Sr/^{86}Sr$ or as Nd 120". Pb isotopic compositions are recalculated using the A notation of Hart (1988), which specifies the extent to which =$^{207}Pb/^{204}Pb$ and $^{208}Pb/^{204}Pb$ are distinct from a reference line defined by oceanic basalts from the northern hemisphere. This notation is used because it is an effective way to identify igneous rocks derived from certain mantle reservoirs such as the DUPAL source. Because U and Th concentrations have not been determined, the Pb isotopic compositions could not be recalculated as initial ratios. However, the isotopic composition of Pb in the samples may still be a useful isotopic fingerprint for the following reasons: (I) relatively little isotopic growth is expected to have occurred; and (2) relatively modest fractionation of U, Th, and Pb is expected during generation of mafic melts from the mantle. The Pb isotopic characteristics of Kohistan samples are similar to the DUPAL mantle reservoir, and this is not likely to be a result of post-eruption radiogenic growth.

The investigated Kamila samples are all mafic, with a restricted range in silica (48-52%) but with a much wider range in incompatible elements such as K. The subdivision of the Kamila metavolcanic amphibolites into E-type (relatively high Ti and other HFSE such as Zr, Y, and Nd; relatively low K_2O, Rb, and Sr) and D-type is supported by trace element ratios (E-type has high K/Rb=660 1330 and low Sr/Nd=8 13 characteristic of MORB and ocean island basalt (OIB), whereas D-type has low K/Rb= 330470 and high Sr/Nd = 18 46, similar to that of modern convergent margin volcanic suites). All of these features are consistent with E-type Kamila metavolcanic amphibolites having a MORB-like basaltic precursor, whereas D-type Kamila metavolcanic amphibolites have clear characteristics of supra-subduction zone igneous rocks. We will show that Kamila D-type metavolcanics are similar in many respects to the rocks of the Chilas plutonic complex.

The isotopic data also distinguish between the two types of Kamila metavolcanic amphibolites. E-type samples have more radiogenic Nd (õNd_{120} Ma=+7.0 to +7.4) and less radiogenic initial Sr ($^{87}Sr/^{86}Sr$=0.7036-

0.7039) than the three D-type samples (õNd_{120} Ma = +2.9 to +4.7; $^{87}Sr/^{86}Sr$Sr=0.7041-0.7047). E-type and D-type amphibolites plot in distinct locations in a plot of Sr-Nd isotopes. The Pb isotopic compositions of E- and D-type metavolcanics are similar. Kamila E-type amphibolites have the most radiogenic Nd and least radiogenic Sr and $^{206}Pb/^{204}Pb$ of all the Kohistan samples that we have analysed. Kamila D-type metavolcanic amphibolites are similar to Chilas samples in having relatively low õNd_{120} Ma and high $^{87}Sr/^{86}Sr$, although only one of the D-type Kamila samples (A125) has a similar Pb isotopic composition. A puzzling feature is that Kamila amphibolites have homogeneous and relatively non-radiogenic $^{206}Pb/^{204}Pb$ and $^{208}Pb/^{204}Pb$, but have variable $^{207}Pb/^{204}Pb$.

Of the four Chilas samples that were analysed, two are from the Main Gabbronorites and two are from the UMA. UMA samples are cumulates, so it is difficult to infer much regarding their petrotectonic setting from their chemical compositions. The gabbronorite samples on the other hand may approximate melt compositions; they are mafic to intermediate and belong to a low-K to moderate K suite. Gabbronorites have high K/Rb and Sr/Nd; we can infer from their Sm and Nd abundances that these were moderately enriched in LREE. The isotopic composition of the four Chilas samples is very homogeneous, with $^{87}Sr/^{86}Sr$i=0.70396 - 0.70414, õ$Nd_{(120)}$=+2.8 to +4.0, and very homogenous Pb. Along with one sample of D-type Kamila amphibolite (A125), the Chilas complex defines the relatively 'enriched' endmember of the Kohistan intraoceanic arc terrane. The Chilas Complex samples and some D-type Kamila amphibolites extend towards an EMI isotopic reservoir in Sr-Nd isotopic space, but extend towards EMII in Pb isotopic space. The Chalt Volcanics define a compositionally heterogeneous group. They include low-K and medium-K rocks of basaltic, boninitic, and rhyodacitic composition. The suite is characterised by very low TiO_2 and Zr contents (<0.45% and <80 ppm, respectively). Chalt boninites and low-Ti basalts may contain very high contents of MgO (up to 12.5%), Ni (up to 400 ppm), and Cr (up to 1500 ppm). Boninitic samples (PL28, N138, N161, and IK580) have 10 to 12.5% CaO, and must therefore be classified as 'High Ca boninites' according to Crawford et al. (1989). Basalts and boninites have very low abundances of Sr (< 100 ppm), Ba (<40 ppm), and Nd (<8 ppm).

Samples of the Chat Volcanics that we analysed have homogeneous isotopic compositions of Nd (õ$Nd_{(120 Ma)}$= +6.3 to +7.3) and Pb (6 $^{206}Pb/^{204}Pb$=18.1 18.4 Ä ^{208}Pb=44-58; Ä ^{207}Pb=5 10). The 8 samples of Chalt boninite, basalt, and rhyodacite fall between Kamila and Chilas suites on plots of Pb isotopic compositions. The Chalt Volcanics define a much larger range on a plot of Sr v. Nd isotopic compositions, reflecting the larger range in H= $^{87}Sr/^{86}Sri$ (0.7040 0.7055). The Chalt Volcanics are metamorphosed submarine lavas, probably reflecting partial re-equilibration accompanying submarine hydrothermal metamorphism. Such metamorphism greatly affects Sr, but not Nd or Pb isotopic compositions. If so, the Chalt before being metamorphosed may have had as restricted a range of Sr isotopic compositions as it now has Nd and Pb isotopic compositions. The petrologic diversity of the Chalt Volcanics can be explained by a fractional crystallisation process from a parental magma similar to the Chalt boninites or basalts. Nd and Pb isotopic homogeneity for the large range of SiO^sub 2^contents in Chalt Volcanics supports this model, although this requires that the range in Sr isotopic composition reflects alteration. Alternatively, the isotopic data are consistent with a model of anatexis whereby Chalt boninites or basalts were melted to generate felsic lavas. It is also noteworthy that the Sr and Nd isotopic composition of the 102 +/- 12 Ma Matum Das tonalite, which crops out along the Hunza River in the belt of Chalt Volcanics, has Sr and Nd isotopic compositions that indicate close petrogenetic affinities with the Chalt.

The range of $^{87}Sr/^{86}Sr$ for igneous rocks of the intra-oceanic Kohistan arc (0.7036-0.7056; mean=0.7043) is similar to that for initial ratios for correlative units from Ladakh, including the Dras Volcanics (c. 0.7035), Kargil mafic cumulates (0.704-0.7053), and Ladakh intrusives (0.7034-0.7048).

The data presented here allow us to address three questions about the Kohistan intra-oceanic arc system. These are: (1) polarity of the intra-oceanic Kohistan arc; (2) nature of the source region of Kohistan arc magmas; (3) palaeogeographic setting of the Kohistan intra-oceanic arc. These items are discussed further below.

Polarity of the Intra-oceanic Kohistan Arc

The intra-oceanic Kohistan-Ladakh arc formed over a subduction zone which dipped beneath it either to the south or to the north. From the time it was first recognised as an intraoceanic arc complex, investigators have preferred the interpretation that the dip was to the north. We are not convinced by the evidence for a northerly dip direction for the Kohistan subduction zone prior to collision with the Karakorum. Regardless of the orientation of the subduction zone beneath Kohistan, another northdipping subduction zone beneath Karakorum is required to generate c. 95 Ma granodioritic rocks of the Karakorum batholith. Collision between Kohistan and Karakorum to form the Shyok Suture is constrained to have occurred sometime in the early Late Cretaceous, 85 102 Ma ago. Pudsey (1986) argued that this suture represented a collapsed back-arc basin in part to accommodate the lack of evidence in the suture melange for subduction of a large oceanic tract.

Our understanding of the distribution of igneous rocks and their trace element and isotopic compositions in Kohistan presents another perspective for constraining subduction polarity of the intra-oceanic Kohistan arc. We conclude that the chemical and isotopic composition of Kohistan igneous rocks, particularly the Chalt and Kamila volcanic sequences, formed over a south-dipping subduction zone. We appreciate that our interpretation is not exclusive: for example, Treloar et al. (1996) infer that the Kamila provides substrate to Chalt cover. However, we find that the geochemical and isotopic data are more consistent with Kohistan sequences reflecting changes across an approximately contemporaneous arc system. There are two lines of evidence to support this. The Chalt Volcanics (especially their boninitic component) are interpreted to have formed in a forearc setting.

There are several models for why boninites form in forearcs, including subduction initiation, spreading-centre subduction, and interaction of the elevated thermal regime associated with a back-arc basin spreading centre superimposed on the high water flux of a forearc. Discussion continues regarding which of these models is most appropriate but the consensus that boninites form in fore-arc settings remains. The interbedded submarine boninites, low-Ti tholeiites and felsic rocks of the Chalt Volcanics are reminiscent of the Izu Bonin Mariana fore arc as is the rapid transition from volcanism to carbonate-dominated pelagic and hemipelagic sedimentation represented by the Chalt-Yasin succession.

The MORB-like affinities of Kamila E-type metavolcanic amphibolites and their intercalation with D-type amphibolites which have a strong subduction component is most consistent with the interpretation that these formed in a back-arc basin. MORB-like magmas are only found at convergent margins in a back-arc basin setting, and sequences containing alternating or interfingering MORB- and arc-tholeiites are characteristic of modern back-arc basins such as the Mariana Trough and Lau Basin.

The inference that the 'subduction component' becomes less important from the Cha.lt Volcanics in the north to the Kamila Amphibolites in the south is also supported by an overall increase in the incompatible HFS elements Ti and Zr in the same direction. Abundances of HFS elements in an intraoceanic convergent margin increase from the forearc to the back-arc, which reflects an overall decrease in the degree of melting or degree of source region depletion that gives a clear facing direction for the arc. The HFSE evidence for Kohistan is very clear; considering only rocks with 48-59% SiO_2 the Chalt Volcanics contain <0.3% TiO_2 and <30 ppm Zr; Chilas Complex samples contain 0.84 and 1.24% TiO_2 and 71-133 ppm Zr; Kamila samples contain as much as 2.25% TiO_2 and 149 ppm Zr.

This ocean closed as a result of two subduction zones one dipping north beneath Karakorum and the other dipping south beneath Kohistan. This tectonic situation has a modern analogue in the Molucca Sea, south of the Philippines. The Molucca Sea is closing as a result of west-dipping subduction beneath the Sangihe Arc to the west and east-dipping subduction beneath the Halmahera Arc to the east.

There is no evidence for a hiatus in India's northward motion during the Late Cretaceous, and continuous northward motion requires a new subduction zone to form immediately after Kohistan and Karakorum collided. This new subduction zone dipped north on the south side of Kohistan. This reversal of subduction polarity beneath Kohistan during the Late Cretaceous is analogous to the situation inferred for the Solomon Arc during the Late Tertiary. Thickened crust of the Ontong Java Plateau arrived at the Vitiaz Trench on the north side of the Solomon Arc during the Miocene. This stopped south-dipping subduction and led to the formation of a new, north-dipping subduction zone southwest of the Solomons. Our model satisfies all tectonic constraints, including: (1) India began moving north during the magnetic quiet zone, at approximately 120 Ma, and continued to move rapidly north until collision with Asia about 50 Ma ago. The northward movement of India reflects the development of subduction zones to the north at about 120 Ma. (2) North-dipping subduction is required beneath Karakorum to generate the Karakorum batholith. (3) Following the formation of the Shyok suture, the subduction zone associated with the northward motion of India dipped north on the south side of accreted Kohistan. This explanation is also consistent with the presence of S-dipping structures in northern Kohistan, which may have formed in association with a S-dipping subduction zone.

Nature of the Source Region of Kohistan Intra-oceanic Arc Magmas

One of the important controversies surrounding convergent margin magma-genesis concerns the relative contributions provided by mantle as opposed to subducted crust and sediments. Understanding this chemical budget is important not only for understanding arc magma-genesis, but also for understanding a wide range of other first-order questions about the earth, including whether or not the Earth's mantle and crust are still differentiating, whether the continents are growing or shrinking, and how the sources of hot-spot mantle formed. It is widely acknowledged that a hydrous subduction zone environment may be responsible for fractionating elements that are not fractionated in anhydrous magmagenetic environments such as hot spots or mid-ocean ridges. The most spectacular of these incompatible element fractionations in arcs is the relatively high abundance of fluid-mobile alkali metals and alkaline earths (K, Rb, Ba, Sr) relative to the abundance of less mobile high field strength elements (e.g., Ti, Zr, Y, Nb, Ta). Having acknowledged these fractionations and the fact that some elements are clearly derived from subducted components such as ^{10}Be, the question remains as to the relative importance of contributions of other incompatible elements that are extracted from the mantle compared to those that are recycled from subducted crust and sediments.

We argue that the distinctive isotopic signature of Indian Ocean mantle is readily recognised in the composition of Kohistan igneous rocks, and that this signature overwhelms any signature from subducted sediment. Hart (1984) recognised a zone of enriched mantle underlying equatorial oceanic regions which he called DUPAL. DUPAL mantle generates lavas with distinctive isotopic characteristics, including relatively low eNd and $^{206}Pb/^{204}Pb$, and relatively high $^{87}Sr/^{86}Sr$ and Delta8/4. Regions underlain by DUPAL mantle are presently far to the south of Kohistan, but DUPAL isotopic characteristics are readily identified in all igneous rocks from the intraoceanic phase of the Kohistan arc.

Another perspective is gained from comparison of Kohistan igneous rocks with Deccan traps. The Deccan Traps are flood basalts that erupted about 65 Ma when India passed over the Reunion hotspot. Most of the Deccan basalts have been significantly contaminated by continental lithosphere, but the Ambenali unit in the upper part of the sequence is thought to manifest predominantly asthenospheric isotopic compositions. The Ambenali basalts formed about 3500 km south of their present position and thus represent the composition of the asthenosphere beneath the subequatorial Indian Ocean basin at the end of Cretaceous time. They show DUPAL isotopic characteristics, especially high delta 8/4, and constitute an appropriate data set that can be used to examine the potential mantle controls on the isotopic composition of Kohistan igneous rocks.

Six samples of Ambenali basalts analysed by Lightfoot et al. (1990) have eNd of +4.3 to +7.5 and $^{87}Sr/^{86}Sr$ of 0.70390.7046, very similar to the range of Kohistan igneous rocks (õNd= +2.8 to +7.4 and $^{87}Sr/^{86}Sr$=0.7036-0.7055). Differences between the Sr and Nd isotopic systematics of Kohistan igneous rocks and Ambenali basalts consist of extension of the Kohistan suite to slightly lower õNd that nevertheless lie within the mantle array, and the higher $^{87}Sr/^{86}Sr$ of the Chalt Volcanics. The Pb isotopic compositions of Kohistan igneous rocks are also similar to those of Ambenali basalts.

The Ambenali basalts have a somewhat lower range of $^{206}Pb/^{204}Pb$ (17.7-18.3) than Kohistan rocks (17.9 18.6), but are similar in lying above the northern hemisphere reference line (NHRL) in similarly distinctive fashions. On a plot of $^{207}Pb/^{204}Pb$ v. $^{206}Pb/^{204}Pb$, both define a trend that obliquely intersects NHRL at $^{206}Pb/^{204}Pb$ 17.6 and extends towards the EMII reservoir. On a plot of $^{208}Pb/^{204}pb$ v. $^{206}Pb/^{204}Pb$, both define a trend which lies above and parallel to NHRL. The Pb isotopic characteristics for both Kohistan and Ambenali are consistent with mixing between mantle reservoirs EMI and EMIT. Intriguingly, evidence of such mixing cannot be inferred from Sr-Nd systematics.

Palaeogeographic Setting of the Kohistan Intra-oceanic Arc

Our inferences regarding polarity of the intra-oceanic Kohistan arc and its formation in an oceanic realm dominated by DUPAL mantle sources lead us to propose a palaeogeographic model for Early Cretaceous time. It shows Kohistan lying within the DUPAL region, about 3000 km south of the Karakorum. The Shyok Sea, that part of Neotethys north of Kohistan, lies between Kohistan and Karakorum. India and Karakorum move northwards together until Kohistan collides with Karakorum, when a new, north dipping subduction zone is constructed on the south flank of Kohistan. The development of this new subduction zone may be manifested by the first stages in the development of the Oman subduction zone (obduction of the Semail ophiolite), about 90-95 Ma.

This model is subject to a number of uncertainties which affect the size of the seaway that lay between Kohistan and southern Asia. First of all, the location of the southern margin of Asia at this time is controversial. For example, we present the configuration of Scotese et at. (1988) which places it about l SoN, but Searle (1991) places it closer to 10 degN. Secondly, the location of Kohistan is not highly constrained. It must lie between India and Asia, and we infer from the isotopic data that it lay in the DUPAL realm, but where precisely? We have also assumed that the surface projection and location of the DUPAL anomaly has not changed significantly in the past 100 million years. These uncertainties indicate that we cannot use the isotopic data to quantitatively reconstruct Kohistan's paleolatitude in Cretaceous time; nevertheless it should be noted that the paleomagnetic study of Yoshida et al. (1996) place Kohistan slightly south of the Equator about 100 Ma ago.

Another way to look at these uncertainties is to try to minimise the size of the Shyok Sea. If we accept Searle's (1991) position for southern Asia at 10 degN and a location for Kohistan closer to the northern limits of the DUPAL anomaly, perhaps along the equator, the Shyok Seaway shrinks in width to perhaps 1000 km. There is nothing in our data that precludes alternate configurations such as this. Geochemical and isotopic data reported here are most consistent with the model that the Kohistan intra-oceanic arc faced north, with forearc crust in the north (Chalt volcanics) and backarc crust in the south (Kamila amphibolites). The isotopic data indicate Kohistan igneous rocks formed by melting of a mantle source with a strong DUPAL isotopic signature. Such mantle sources exist in the oceanic realm far to the south of the present tectonic setting of Kohistan.

Bibliography

Adamson, A. W; Gast, A. P.: *Physical Chemistry of Surfaces*, Wiley-Interscience; New York, 1997.

Adkins, C.J.: *Equilibrium Thermodynamics,* Cambridge: Cambridge University Press, 1983.

Balescu, Radu: *Statistical Dynamics: Matter Out of Equilibrium*, Imperial College Press, London, 1997.

Balkwill, Dr. Fran: *DNA Is Here to Stay*, Carolrhoda Books, Inc., Minneapolis, 1993.

Campbell, I. M.: *Introduction to Synthetic Polymers, USA*: Oxford University Press, 2002.

Clark, M. M.: *Transport Modeling for Environmental Engineers and Scientists,* Wiley, New York, 1996.

Crompton, T. K.: *Analysis of Polymers*: America: Pergamon Press, 2005.

David Baraff: *Dynamic Simulation of Non-Penetrating Rigid Bodies*, Cornell University, Ithaca, NY, 1992.

Duxbury, G.: *Infrared Vibration-Rotation Spectroscopy, From Free Radicals to the Infrared Sky*, Wiley, NJ, 1999.

Esumi, K.; Ueno, M.: *Structure-Performance Relationships in Surfactants*, Marcel Dekker, New York, 2003.

Folk, R.L.: *The Petrology of Sedimentary Rocks*, Hemphill Publishing Co., Austin, 2001.

Fritz Helmet: *Diffraction in Crystals*, Ivy Publishing House, Delhi, 2011.

Gurugnanam, B. : *Essentials of Hydrogeology*, New India Pub, Delhi, 2009.

Harlyn Baker, H. : *Building Surfaces of Evolution: The Weaving Wall*, International Journal of Computer Vision, 1989.

Herbert, N.: *Quantum Reality: Beyond the New Physics*, New York: Doubleday, 2003.

Holmgren, D. A., Moody, J. D. : *The Structural Setting for Giant oil and Gas Fields*, Tokyo, 1975.

Israelachvili, J.: *Intermolecular and Surface Forces: With Applications to Colloidal and Biological Systems*, Academic Press, San Diego, CA, 1992.

Jackson, J.D.: *Classical Electrodynamics*, John Wiley and Sons, New York, 1975.

Keller, G.: *Separations: New Directions for an Old Field*, AIChE, New York, 1987.

Kurt Georg: *Numerical Continuation Methods, An Introduction*, Springer-Verlag, New York, 1990.

Lifschitz, E. M.: *Fluid Mechanics*, Pergamon Press, North Holland, 1987.

Lindon, John C.: *Encyclopedia of Spectroscopy and Specrometry*, Academic Press, Uk, 1999.

Majumdar P.K. Basu: *Law of Mines and Minerals*, Universal Law, Delhi, 2011.

Miller, J. M.: *Separation Methods in Chemical Analysis,* John Wiley and Sons, New York, 1975.

Neil Gershenfeld: *The Physics of Information Technology,* Cambridge Univ. Press, Cambridge, 2000.

Olagoke Olabisi: *Handbook of Thermoplastics,* Marcel Dekker, New York, 1997.

Prutton, Martin: 1994 *Introduction to Surface Physics.* Oxford University Press. 1994.

Seidel, A.: *Kirk-Othmer Encyclopedia of Chemical Technology*, Wiley-Interscience, New York, 2004.

Simeonov, V.: *Information Aspects in Chemical Analysis*, Obrazovatelni Tekhnologii, Sofia, 1999.

Tanford, C.: *The Hydrophobic Effect. Formation of Micelles and Biological Membranes*, J. Wiley, New York, 1980.

Walas, S. M.: *Phase Equilibria in Chemical Engineering,* Butterworths, Reading, MA, 1985.

Whitmore, G : *Chemical Crystallography And Lipid Crystals*, Sarup Book, Delhi, 2009.

Index

❑❑❑